# THE LAST
# SUPPER

# THE LAST SUPPER

## HOW TO OVERCOME
## THE COMING FOOD CRISIS

# SAM KASS

CROWN
NEW YORK

CROWN
An imprint of the Crown Publishing Group
A division of Penguin Random House LLC
1745 Broadway
New York, NY 10019
crownpublishing.com
penguinrandomhouse.com

Library of Congress Cataloging-in-Publication Data has been applied for.

Hardcover ISBN 978-0-451-49496-2
Ebook ISBN 978-0-451-49498-6

Editor: Francis Lam
Editorial assistant: Darian Keels
Production editor: Craig Adams
Text designer: Andrea Lau
Production: Christopher Andrus
Proofreaders: Muriel Jorgensen and Chris Jerome
Indexer: J S Editorial, LLC
Marketer: Mason Eng

Manufactured in the United States of America

9 8 7 6 5 4 3 2 1

First Edition

The authorized representative in the EU for product safety and compliance is Penguin Random House Ireland, Morrison Chambers, 32 Nassau Street, Dublin D02 YH68, Ireland, https://eu-contact.penguin.ie.

To Rafael Thiha Kass, Cy Mindon Kass, and all future generations—
may we rise to the responsibility of
leaving them a world worth inheriting

# CONTENTS

# PART I

## SETTING THE TABLE

# A Mindful Meal

"I'm screwed."

That is a polite way of expressing my thoughts upon meeting the chef with whom I was supposed to prepare a banquet at the World Economic Forum.

Nearly twenty-five hundred dignitaries from 140 countries had assembled in Davos, Switzerland, for the annual networking extravaganza. Their numbers included no fewer than forty heads of state. Among them: David Cameron of the U.K., Justin Trudeau of Canada, and Jacob Zuma of South Africa. Ban Ki-moon, John Kerry, and Joe Biden were in town, as were Mary Barra, the head of General Motors, and Satya Nadella, of Microsoft. Yo-Yo Ma, Bono, and Leonardo DiCaprio represented the arts.

And then there was me. My job was to conceptualize and oversee a luncheon for fifty of these luminaries—with the help of a woozy codger standing in front of me.

Anyone who has worked in a professional kitchen recognizes the type. He had thinning gray hair and looked to be well into his sixties, an age when even the most physically fit cooks find it tough to summon the energy to prepare a banquet. He had the telltale puffy face, watery eyes, and flaming complexion of someone who habitually uncorked a bottle of red wine upon rising in the morning and guzzled the dregs of his second or third bottle just before going to bed at night. I surmised that like most of his kind, he cooked dated French catering fare—bacon-wrapped green beans, beef filet with an insipid demi-glace, gummy potatoes au gratin—food he considered good, or rather good enough for the guests at a faded hotel in a Swiss town with a permanent population of just eleven thousand, where the main draw is powdery snow and well-groomed ski runs, not haute cuisine. He seemed a relic of the past. My meal was supposed to be all about food's future—or, more accurately, its lack thereof.

I had come to Davos in 2016 at the request of the United Nations' World Food Programme. They asked me to put on what I call a Last Supper, a concept I developed as a way to demonstrate how climate change will affect what we eat. "Demonstrate" is the key word. The six years that I cooked nightly dinners in the White House for the Obama family, while also serving in an official capacity as the administration's senior nutrition policy adviser, taught me valuable lessons about the unrecognized power of a meal and also about how leadership works in the real world.

You can lecture all you want about the perils of global warming, and listeners' eyes will glaze over. Repeating hackneyed statistics like "The world will soon become two degrees warmer than it is now" doesn't motivate people at all. Hell, as someone born and raised in Chicago, I guarantee that many residents of my hometown would say, "Two degrees warmer? Great! I'll take it, let's warm this place up!" Especially on a January morning with a north wind howling off Lake Michigan. Beyond the fact that it doesn't sound like a serious problem, no one connects in an emotional way to two degrees. Any marketer will tell you that if you want someone to change their behavior, it is imperative that you create an emotional connection with them. One of the biggest failures of the climate and environmental movement has been its inability to connect with people on that level.

A good example is straws. As bad and unnecessary as plastic straw waste is, straws make up a statistically insignificant amount of plastic that we put into the world. But surely you've seen your local restaurant or café act as the environmental battleground on which the war on straws is being fought. Why do we care about it? Because we see those horrible pictures of sea turtles with straws stuck in their noses and they pull on our heartstrings. We feel empathy and guilt for the turtle, sadness that it has to live with *our* plastic lodged in *its* body. And now we see paper and seaweed and bamboo straws all over, Starbucks takes them out of their stores, and so on. So the question I asked myself was, How could a meal, one that leaders literally consume, connect them to the issues and the stakes and change their attitudes as a result?

You'd like to believe that people make decisions rationally, based on evidence and science. You'd really like to believe that people in high offices, where they have access to advisers who know the evidence and science, especially do so. But people, even people in high office, make important changes based on *feelings*, not just cognitive analysis. I've spent the kind of time in Washington that offered me access to power boardrooms and government offices. And I saw this firsthand over and over again.

Sometimes the data is just so strong it's a clear choice. But more often than not, choices being made in these rooms were full of difficult trade-offs and risks, political and otherwise. Remember that data is not a magical, objective cure-all. The type of data presented in a meeting, or the way it's presented, is subject to human decisions. So data is often just one part of the puzzle. How decision makers feel— call it their gut, or call it their intuition if that sounds better than their "feelings"—inevitably has a role to play. Time and time again I have watched decision-makers—Barack Obama included—find the resolve to make the tough but right decision even at a high cost when they connected with the issue on a deeper level and understood the real-life implications.

At my suppers, I would create a memorable menu—full of many of the ingredients that have brought joy to our lives and that we eat on a daily basis. As people began to enjoy their first bites, I would stand up to speak. But instead of rambling along with the usual "cheffy" spiel about the local artisans who made the cheese or the idyllic life the chicken lived, I'd announce, "Welcome to the Last Supper."

There would usually be a rumble of uneasy laughter as diners wondered, "What the hell is this?" Then I'd say, "Experts are increasingly certain that our children and grandchildren will not be able to taste many of the items you enjoyed at this meal."

I'd raise a glass and point out that because of climate change the wine they sipped may no longer be available or, if available, not be of the quality they expect and at a price they can afford. To produce decent wines, grapevines need consistent weather. Sudden swings in temperature or periods of excess rain—or drought—ruin a vintage. Ruinart is one of the oldest and most prestigious champagne producers in France. It has been making champagne since 1729 using the same single-grape variety for generations. In 2023, for the first time, it released its champagne with a blend of different grapes: It could not produce enough of the time-tested variety. It's getting so bad that vineyard owners in Champagne are buying land in Britain as a hedge against the day when weather patterns will make the English Midlands more amenable for producing fine sparkling white wine than the legendary French province. You *know* it's bad when the French are buying land in England to make champagne! Indeed, the National Academy of Sciences predicts if temperatures rise by two degrees, which will likely happen in the next few decades, the world's viable wine-growing regions will be cut by *more than half.*

After I imparted that information, the room would fall completely silent.

Having caught my guests' attention, I'd urge them to savor the chocolate in the dessert. Cacao, from which chocolate is made, demands the warm, wet, and above all stable conditions found within about 10 degrees latitude of the equator, primarily in South America and Africa. Basically all of the chocolate the world eats is produced by smallholder farmers in those regions. Assuming we hit the two-degree threshold, hotter and drier weather will leave nearly all of cacao's current growing regions unsuitable for the crop.

The startled look would spread to more in the room.

I'd ask how many people had a cup of coffee that morning. Nearly the whole room's hands would go up. Raise your hand if you had two cups. Most keep their hands raised. Three cups, fewer hands but still many. Four cups? I tell those whose hands are still up that I am worried about them and we should talk after the dinner. But coffee, which for about a billion people is somewhere between a daily pleasure and a necessary drug, is also one of the most sensitive crops grown. The best beans are raised at high altitudes often at the base of mountains in warm—but not *too* warm—regions. Hotter weather is forcing growers to seek cooler temperatures ever higher on mountainsides and to abandon lower-level plantings. Research shows that because of climate change *two-thirds* of the world's wild coffee species are threatened with extinction. You may not be drinking liquid extracted from the beans of these undomesticated plants in your morning cup of joe, but they are the gene bank we need to keep growing coffee. Quality arabica coffees survive today only because they have been crossbred with wild cousins that have immunity to otherwise fatal diseases.

Wine, chocolate, coffee. I'd add tea, because tea's plight is the same. By this point, basically everyone in the room has that look on their face. Depending on the menu, I might add that oysters, lobsters, clams, pistachios, and almonds are among foods that we should enjoy while they last.

This same story is playing out for many, many other vitally important and beloved foods. Something so closely associated with a geographic area as Georgia peaches is also in trouble. José Chaparro, a geneticist specializing in stone fruit at the University of Florida, predicts that the peach industry in Georgia and other southern states could collapse due to erratic weather. He is urgently trying to breed tolerant varieties, but for Georgia peach farmers the future is here. In 2023, more than 90 percent of their crop was lost due to a heat wave and then early frost. We will probably still have peaches, but there will be a big disruption on where we grow them when farmers are forced to replant orchards in more agreeable regions—at tremendous cost.

Even at the highest levels of policymaking, food is a powerful communication tool. Everybody relates to it. And because of climate change, many of the foods we love are in real danger. But the good news is, food can also have a huge role in how we mitigate climate change, if we get enough people to take action.

To drive home the link between what we eat and the changing climate, for my Last Supper events I always serve menus that include threatened products from the regions hosting the gatherings, places often thousands of miles away from where I live. Having created these meals on three continents, I've learned that it is critical to partner with local chefs who know their areas' native ingredients well and will choose the ones they're best at preparing. I need my colleagues to become personally invested in the Last Supper concept. And I really didn't think the arranged marriage between me and the chef in Davos was going to work out.

## A POTENTIAL CALAMITY

In defense of my initial reaction to my Davos kitchen partner, putting your name and reputation in the hands of someone you have never met is always fraught. I had already experienced a few near-disastrous Last Suppers, most notably one at the U.S. ambassador's mansion in Paris for the 2015 UN Climate Change Conference (aka COP 21), where the young, inexperienced crew I worked with nearly failed to get the food to the tables at all. After realizing they were floundering, I found myself drenched in sweat over a stove trying to get 130 plates of pasta out before putting my suit jacket back on to go discuss the future of wheat. I was well aware that being the American First Family's chef had put me in a position that could bring glowing successes—or debilitating, highly visible failures.

The skills I initially mastered as a young man in restaurant kitchens have gotten me into a lot of rooms that are hard to enter and would have otherwise been strictly off-limits to someone in his thirties with a BA in history. And being able to get into those rooms, for whatever reason, made me feel a sense of responsibility to represent people and families from around the world who are grappling with the consequences of unhealthy food, too little food, and climate change. "Maybe I don't belong here, but now that I'm here," I tell myself, "I better make the best of it. Sam, this meal has to resonate. DO NOT SCREW IT UP!"

The near crisis at the ambassador's mansion came rushing back as the Davos chef approached me in his empty kitchen, extended his hand, and began speaking. Motioning me toward the stoves, he made me understand through a series of broken phrases and pantomime that he'd thought a lot about this event and that he'd like to talk over the menu.

As he worked over the stove, he haltingly described each

ingredient and what was happening in the region because of climate change. The main course would feature saibling, a relative of Arctic char and North American brook trout. It is an iconic species in the area. In her famous 1939 essay, "Three Swiss Inns," M. F. K. Fisher described trout from Swiss mountain streams as the "most delicious fish" she had ever eaten. Sadly, the chef informed me, warming water in the streams had killed most of those fish. The ones that survive suffered from poor reproduction rates. A risotto dish to accompany the saibling was made from locally grown rice. The chef told me that when he was a kid, they used to grow a lot of rice in the area, but because of a lack of water, now they grew hardly any.

The Davos chef's knowledge and depth of passion about the ingredients he had chosen surprised me, but shouldn't have. Of course, a lot of chefs—whether working in high-end restaurants or in worse-for-the-wear ski lodges—care about their ingredients, care about where they come from, and increasingly care about the men and women who produce them and the land and water from which they came. If you are a professional cook and find those things meaningful, inevitably you are encountering challenges brought about by climate change.

I felt like an ass for writing my Swiss colleague off as a mere old-school hotel chef. He had gotten completely absorbed in the Last Supper message and done a lot of research to assemble a thoughtful menu. His cooking told a great story, one that meant a lot to him, and he was excited and proud to share it with world leaders. I savored a quiet smile of relief. Observations about the universality of food have become hackneyed, but their veracity couldn't have been clearer to me than it was in that cramped kitchen in that weird town in the middle of the mountains at that otherworldly event.

Unlike my Parisian crew, this chef also knew how to feed a crowd fast and well. To this day, I have no idea how he got fifty servings of trout on the tables at nearly the same time, all crispy-skinned and cooked to a beautiful medium rare. Can you remember the last time you sat down to a large dinner where the fish was cooked perfectly?

The luncheon was an ideal prelude to the post-meal remarks delivered by Christiana Figueres, then the executive secretary of the United Nations Framework Convention on Climate Change. Often described as a force of nature (an understatement if there ever was one), Figueres is one of the few people who felt like a real rock star to me after six years in Washington. (Call me a climate nerd, that's fine!) In 2010, following the implosion of negotiations toward a global climate treaty in Copenhagen, Figueres was tasked with the challenge of breaking through the paralysis. "Save the planet" was the job description given to her. "You have full responsibility, but absolutely no authority."

By 2015, she had persuaded (with a lot of support and muscle from President Obama) 195 governments to sign the Paris Agreement and change the course of their policies toward building stable economies that did not depend on enormous carbon emissions. From her position of near-universal respect, Figueres drove home the Last Supper's message with more authority than I ever could have and with a greater sense of urgency. "You all need to start making a difference today. As you just heard, the stakes are high, and we are out of time," she told the by-then well-fed dignitaries.

The Davos meal was a tremendous success. But what is really accomplished by one of my Last Suppers? I'm not so naïve as to believe that one lunch, even a very good one, changes the course of world history. But if your goal is to get people to make a connection between food systems and climate change, a powerful way to drive home that message is to serve them a beautiful piece of fish and then inform them that the delicacy they just enjoyed will be gone in a few years. These experiences stick with people. The satiated bigwigs leave the tables familiar with the issues and knowing, in a personal way, what is at stake. They have a connection.

At the 2015 UN Climate Conference in Paris, where the famous Paris Agreement was hammered out, food systems merited barely a mention during official working sessions. At the 2023 conference, for the first time ever, an entire day was devoted to agriculture, 130 countries signed on to a declaration promoting sustainable farming, and the final statement acknowledged that sustainable farming was a necessary part of the solution to climate change.

I ran into Christine Lagarde, then the chairwoman of the International Monetary Fund, at a gathering several months after she had attended one of my suppers. She approached me smiling and started reminiscing in great detail about every course I'd served. Lagarde is from France, and I have yet to meet someone from that country who doesn't appreciate—and remember—a great meal. She also acknowledged that she had never before associated gastronomy with global warming.

That encounter might seem like a small success, but we will need every success we can get if our societies are going to survive the coming climate catastrophe and the mass famines that it will cause. We're already beyond the point where small successes, while important, will be enough for our food system to adapt to the changing world.

Tragically, it's already too late to expect that we can stop, let alone reverse, global warming in time. Our challenge now is to focus on actions we can take *now* to slow the rate of climate change in an effort to buy time and to prepare ourselves to deal with the inevitable negative impacts on how we feed humanity.

Our food choices will play a determining role in our success. Climate change's dire effects go far beyond coffee and chocolate, of course. Farmers who raise corn, rice, wheat, and other staples all face the increasing risk of crop failures as a result of floods, droughts, and other adverse weather events the world over. But at the same time, modern agriculture is a major cause for *why* it's becoming harder to farm. Agriculture contributes an estimated *one-third* of the greenhouse gases that are warming the world. Yet—and this is key—farming differently can be our very best lever for mitigating climate change. Food production is a cause of, but can be a cure for, the climate crisis.

## BELLY OF THE BEAST

My own experiences in the White House trying to bring about improvements in America's food system showed me firsthand that even in the halls of power making the most incremental changes in government and commerce at a very large scale is a daunting task. Nothing comes easy. And it generally doesn't come with blueprints or instruction manuals.

Thanks to the work of concerned food writers with best-selling books and prominent newspaper columns, by 2008, I was already well aware of the problems with the way we feed ourselves. That didn't make me special; by then, most conscientious eaters were. A nascent cultural awakening had begun in America led by chefs and farmers and a handful of journalists, authors, advocates, and academics. I was a product of that shift and a student of the writers who had been shaping the dialogue and discourse. Their hard-hitting, well-articulated critiques helped crystallize and elevate the issues we were facing. These kinds of critiques can be the beginning of change. They were, after all, the beginning of my change. People who cared about improving the food produced in America had some tremendous and high-profile voices pointing out what was wrong.

But once in the White House, we soon found the limits of where we actually stood as a community ready to make change. There were leaders, largely people working on their farms, in their kitchens, in churches and community centers. But I came to quickly understand that writers and critics are not necessarily leaders. These voices largely stayed focused on utopian visions of the future untethered to reality. These visions were compelling and largely in line with the future I have been working to create. But they lacked even the semblance of a plan to take us there. They spent most of their time laying out overly simplistic narratives of how broken the system is, telling us the government and its subsidies are the root of why we eat what we do, and if we could only change them, we would be feeding ourselves in the right way. And they railed that the world didn't look like their vision.

Few of the loudest voices addressed the more complex issue: *How* do we actually go about fixing food production? I can't blame them, to be fair. Until Michelle Obama came to the White House, no presidential administration had ever attempted to make extensive improvements to the national diet at the executive level in the modern era, meaning those of us who had shared values around food production had never been on the inside. We had been on the outside yelling at the top of our lungs that we had a problem and something should be done. When she put me in charge of carrying out her vision, it was hard to know where to start, or where making change was possible, and what that change would even look like. No one knew.

The scattershot approach of advocates in the food movement was testimony to this difficulty. Do we ban chemical pesticides? Forbid the development of genetically modified crops? End the abuse of livestock? End hunger? Enact stringent water-quality controls on large, confined hog and chicken operations? Improve food safety? Encourage small farms and organic practices? Eat locally? Patronize farmers markets? Eliminate subsidies to industrial corn and soybean farmers? Ban administering subtherapeutic antibiotics? I could go on. But in an administration, you have limited time, funding, and political capital. Every choice matters. So, where to start?

WE CHOSE TO FOCUS ON improving children's health and nutrition for a number of reasons. First, it was deeply meaningful to the First Lady as a mom who had experienced these issues firsthand with her two girls, even though she was well educated and had resources. Politically, the nexus between food production and health represented a strong position. Fighting for children and families in the face of declining kids' health was hard to argue against. In Washington, that has a big impact on what you will be able to accomplish. You have limited time to move anything through the obstacles put up by bureaucrats, legislators, and lobbyists. Having a position that is hard to refute gives you a chance to get something done. This focuses the mind: What *can* we realistically do? We knew that the vast majority of senators and members of Congress would support—or at least not vociferously oppose—getting better food to the nation's kids. By design, our framing was not threatening. It was substantive but not overly policy focused at its launch, which made people underestimate what we were setting out to do. It allowed us time to distill our strategy and make real progress before some of the country's biggest and most powerful companies and the political opposition really gauged how hard we were working for real change and how they would be able to fight against us.

Finally, the focus on eating more nutritious food provides a twofer: It makes people healthier, and in many cases it helps lower agriculture's contribution to a deteriorating environment. Cutting back on red meat is good for your arteries *and* reduces food's greenhouse gas emissions. Fruits, vegetables, and grains help control blood sugar and cholesterol *and* have a much lighter environmental footprint than foods that make up much of the typical American diet. Given how much time I spend focused on the nexus of food, agriculture, and climate change now, it is interesting to look back on how little that was explicitly discussed when we entered the White House. But it was always on my mind.

Even though I'm now convinced that reducing food production's contribution to climate change is perhaps the most immediate challenge facing humanity, the steps we took in Washington to make healthy food more readily available provide a template of how our food system can be changed in its entirety. Perhaps more important, our efforts also showed some of the limitations of trying to get improvements through the government. If we are going to save the planet from a complete climate breakdown, we will have to look for solutions beyond Washington, D.C.

That said, during the six years I worked in the Obama administration, we managed to succeed in making unprecedented changes. More than thirty million schoolchildren now have access to affordable, nutritious meals in their cafeterias each and every day, including breakfast for all kids who attend low-income schools. We eliminated the once common practice of adding heart-destroying trans fats to processed foods, saving thousands of lives. With our help and prodding, Walmart, with its tremendous buying power, forced reductions in sugar and salt across the industry and widened its selection of healthy foods by selling them at the same affordable prices as less-good-for-you products. Realizing that doing good and increasing profit could go hand in hand, the retail giant subsequently began taking steps to drastically reduce its carbon footprint.

Six years on the front lines taught me about how real change in the food system can come about. This book tells that story. Although my critique of how we produce our food is as harsh as many that have been written before, it is not meant to be just another in a long line of familiar diatribes. Lord knows there has been enough of that, with far too little to show for it. What follows is in part an action plan based on my experiences—a handbook of sorts, with concrete examples that demonstrate what we need to do ourselves, and what we need to encourage government and business to do in areas over which we have limited personal control, to bring about real change and how to go about doing it. To be successful, good-food advocates have to understand how government and business *really* work if we want to get them to change. I will provide examples because, as I learned, change is far more complicated than I had imagined when looking in from the outside. The public's failure to recognize realities is a major impediment to bringing about the necessary improvements in our food system.

This book was shaped by the fundamental challenge Michelle Obama gave me when we first walked into the White House. She said, "Okay, we are here. Now go make change, real change, for the people who need it the most and for as many people as possible. We are not here for show, or praise. I care about making the food people eat better and improving people's lives." In that instant, my time being a young advocate who saw everything through the lens of my purest ideals ended. What emerged was the realization that the only path forward was to grapple with the actual reality we faced, and in all that complexity start to forge a path toward my ideals. It is a profoundly different orientation to the same problems and analysis. We have to meet reality where it is and strive for big bold change from that reality. There is literally no other way.

This book is for those who are serious about making real change to affect as many people as possible. Many of my most progressive friends and close allies would describe similar visions of how we would ideally want the world to be, and I share those values. I try to live those values in the choices I make. I buy my chicken and pork from a family of farmers I've known for years. Their animals live on pasture and are raised in a regenerative system that is rooted in the health of the soil. They are cared for with the highest welfare practices. It is picturesque, and is how I wish all the world's farms would be. My whole chicken also costs $40.

While I love that chicken and believe it tastes better and is better for the planet, this is not a model that will work to feed the world today. I believe in the small operations that have found ways to most purely express the values I hold; they show us what may one day be possible. But I ground my thinking in the reality of how we currently feed ourselves. That means grappling with the questions of what shifts have to happen to get the big companies that are producing our food to change. I don't think most of them are going away anytime soon. If we are serious, we need to get in the game where the game is actually being played.

# Four Ways to Save Ourselves (Maybe)

My earliest challenge in Washington was to develop an overall strategy of where we wanted to go. And then we had to implement the tactics necessary to get there. So, I broke the job we had to do into four main steps, which I think of as the pillars upon which change in the new food system must be built.

First of all, we had to change American values and attitude toward food. In other words, we would have to change our culture. As I would learn, unless the culture is firmly behind your cause, making progress in any area is tough, and in politics, virtually impossible. This is even truer for business, where executives won't dare move unless they feel the pressure of their customers behind them. But as soon as they do, they will do everything they can to move as fast as possible to chase where the market is going. The foundation of both voters voting and consumers buying is a value system in our culture. To make the greatest progress, *that* has to change to one that says the health of the planet and the health of our communities are top priorities.

Second, we had to then use the shifting attitudes of the culture as a lever to push for changes in government policy and legislation. The major limiting factor to what you can accomplish in

Washington—even aside from ever-present political gridlock—is that the government has limited direct authority to dictate or even directly shape the eating habits of the American public. Think about it. The only people the federal government directly feeds are students and members of the armed forces. In areas where it does have authority, action in Washington is usually painstakingly slow and limited to the areas covered mainly by the Farm Bill, such as SNAP (aka food stamps) and other food assistance programs and through subsidies to farmers—more on that bogeyman later. This is not to say that policy does not have a major role to play; it's just limited in its ability to fundamentally fix the challenges we face.

But businesses, including big businesses—the Walmarts, the McDonald's, the Conagras that form the third pillar—exert enormous influence over what and how Americans eat. At times, they can be as slow and hidebound as government, but when they want to, corporations can implement sweeping change with the stroke of a CEO's pen, no need for interminable negotiations with lobbyists, special interest groups, and opposing politicians. Following the infamous *E. coli* outbreak in California spinach that sickened hundreds of people and killed three in 2006, Walmart immediately imposed field sanitation criteria designed to reduce the chances of a similar disaster recurring. Within a few months, California growers who depended on Walmart (meaning just about every major producer) had all but fallen over themselves erecting animal barriers, installing rodent traps, and clear-cutting bushes and any terrain bordering fields that might harbor a deer, wild pig, mouse, or frog—any potential transporter of *E. coli* bacteria into a field.

Most companies feeding us are selling products that harm our health and contribute mightily to the degradation of the planet. What, then, will need to change to unlock the power and potential for the companies that feed us to become a force for more positive change? If we don't successfully answer that question, I do not believe we will solve the challenges ahead. For all the ills we can justifiably blame big business for, the fact is that we can't just wish them away. We need them to change how they do things with carrots and sticks.

The fourth pillar to bringing about change in the food system is the role new technologies and innovations will play in improving how we feed ourselves. According to estimates, agriculture is responsible for about a third of all greenhouse gases. With adequate investment, new technologies stand poised to rapidly bring about paradigm-shifting alterations to ward off agricultural disaster because of climate change, and promising technologies exist today that can help *reverse* the damage that mass food production causes. Positive innovations should be encouraged financially and supported by an open-minded public.

SO, THAT'S IT. FOUR PILLARS. Tackle them all, and we've got a fighting chance. I'm not going to sugarcoat it—it's not going to be easy. But this book will show how we worked in the administration to change the culture's attitude toward food in concrete ways that paved the way for our legislative accomplishments. Think of Michelle's White House Garden and the media impact of her out there pulling carrots—organic carrots—with inner-city kids. But I'll also talk about how "normal" people, working at the grassroots, hyperlocal level, have managed to effect sweeping change in some of the least "food aware" parts of the country.

I should add that we also had some colossal failures in trying to shape food culture, which I will describe, albeit with a red face. Consider them cautionary tales.

I had to swallow hard before I became comfortable knowing that to make the changes Michelle insisted upon, I'd have to engage with some of the most notorious retailers and restaurant chains out there, the bogeymen of the good-food movement. For better or worse, big food retailers have far more direct influence over what Americans eat than government. The months I spent working with executives from Walmart, the labor movement's nemesis, were particularly taxing for me. My father has dedicated his life to the labor union movement, and I often joined him at union rallies and picket lines when I was young. My ultimate boss, the president, had gotten elected thanks in no small part to organized labor. But persuading the country's biggest food retailer to do better with its grocery offerings brought about immediate improvements in the nutritional profile of its merchandise and lowered the cost of more nutrient-dense foods like fruits, vegetables, and whole grains. Suddenly budget-conscious consumers could afford organic produce and packaged goods with less sugar and salt than conventional offerings. Government could theoretically do much of that, but for myriad reasons we can't rely on it to produce such changes.

After leaving the Obama administration, I launched a new career, helping to create an investment firm called Acre that has a mission to help solve climate change and improve human and environmental health. We invest in and then help build conscientious food and agriculture start-ups. In my new role, I've encountered groups of entrepreneurial scientists who are deploying technological innovation in what can be world-changing ways. The work they do is no longer the stuff of sci-fi or pie-in-the-sky futurism. So much so that agriculture could soon become a major way to *remove* carbon from the atmosphere and put it back in the soil. Some initiatives will enable farmers to profit from "growing carbon," essentially farming in such a way that their land sequesters increasing amounts of carbon each year and then selling "offsets" based on the stored carbon. One promising company deploys microbes to enable common commodity crops to double or even triple the amount of carbon they can pump into the ground.

Another area of focus: meat production. It's one of the biggest contributors to climate change. But a new generation of entrepreneurs is using mycelium (the rootlike structures of mushrooms and other fungi) to create meat substitutes that are totally natural and deliver huge amounts of healthful protein and fiber without the chemical additives central to current plant-based meat replacements produced by the likes of Beyond Meat and Impossible Foods. And unlike Beyond and Impossible, these new products really can have the flavor and texture of a juicy porterhouse steak.

Anyone concerned about climate change should also be excited by—or at least open-minded toward—the products being developed by companies using CRISPR, a gene-editing technology that is faster to develop, a lot cheaper, and much more accurate than conventional genetic engineering. Forget GMOs as you know them, the old scary freak-show examples of glow-in-the-dark chickens with jellyfish genes. That kind of GMO is already on its way out. CRISPR, like all gene editing, is controversial. But applied intelligently, it has the potential to be a game-changing tool in our battle to adjust to hotter, drier, more uncertain conditions. The technology exists today to produce crops that can withstand drought, heat, flooding, and other effects of global warming, no Frankenstein species gene mixing needed. And, to be blunt, we will need to deploy every tool we can to combat the climate catastrophe that is already upon us.

I'm sorry to report that when it comes to climate change, we find ourselves in the position of a wayward college student who has blown off the entire semester and finally has to pull all-nighters in a desperate attempt to pass the final exam. The best that student can hope for is to squeak by and avoid flunking.

It is already too late for humanity to totally sidestep the dire effects climate change will have on every part of our lives, and pointedly on the food we eat. The catastrophe is happening now almost everywhere you look. Even as I write this, in the autumn of 2024, the southern United States has just faced the longest, most extreme heat waves in anyone's memory. Last month, New York City turned into Venice for a day with the hardest rain it's seen in a hundred years, if you don't count the last time it faced rain like that, a few years earlier.

But that only means it's more vital than ever to fight back. We may not *avoid* climate change, but we need to work harder than ever to avoid the *worst* of climate change, which, you don't need me to tell you, can be awful beyond anyone's imagination. This is a fight not to save the planet but to save ourselves. And getting our food right is the first, best, and maybe last hope of that. We have the chance to take the crisis moment we find ourselves in and come out in a far better place than where we started, not just to prevent calamity, but to have a food system that is far closer to bringing the values of good, nourishing food to as many people as possible and ensuring that the people who produce our food are thriving. I believe we have a chance to move far closer to those ideals than I could have imagined at any point since I started cooking all those years ago.

## THE FACTS

Much has been written about the problems we face—detailing the crises in our climate, our food systems, and our health. While this book is focused on what we can and should do next, it's essential to ground ourselves in the scale and nature of the challenges we face. The facts below are not theoretical; they are the realities we must confront if we are serious about creating a food system that is sustainable, resilient, and equitable.

Food and agriculture sit at the heart of both the climate and health crises. The system that feeds us accounts for 21 to 37 percent of global greenhouse gas emissions, and those emissions are rising year after year, according to the Intergovernmental Panel on Climate Change. It is the number one driver of deforestation and biodiversity loss. It uses more than 70 percent of the world's fresh water—a resource that is projected to face a 40 precent shortfall by 2030. That is a figure that was much discussed at the United Nations' UN Water Conference in 2023, and to put it plainly, it means that the world's demand for fresh water will outstrip supply by 40 percent. It's a catastrophic figure.

Food and agriculture are responsible for 78 percent of ocean and freshwater pollution, and these activities currently occupy half of the world's habitable land. At the same time, our oceans absorb 90 percent of excess heat generated by global warming and 25 to 30 percent of human carbon emissions, which is altering marine ecosystems and threatening aquatic food sources.

The planet is losing an estimated 24 billion tons of fertile soil annually, and 33 percent of Earth's soils are already degraded. If current trends persist, 90 percent of soils could be degraded by 2050. One hundred ten billion metric tons of carbon currently in the atmosphere once resided in the soil—equivalent to roughly eighty years of today's emissions. Soil degradation and poor land management are key drivers of this shift; we are releasing more and more carbon into the air not just through factory smokestacks and car exhausts but by tilling it out of the soil, where it once was stored.

Despite the vast environmental footprint of our food system, the diversity of our diets remains narrow. Just three crops—rice, wheat, and corn—make up 60 percent of global calories, and twelve plants and five animal species account for 75 percent of what humans consume. This lack of diversity leaves the system vulnerable to increasing volatility as a result of climate change. Yet, one-third of all food produced is wasted, even as approximately 800 million people are malnourished, 1.4 billion face food insecurity, and 2 billion people are overweight or obese.

In the United States, 89 percent of the food produced by the twenty largest food companies is considered unhealthy. As a result, one in three Americans is overweight or obese, and 136 million people are either diabetic or prediabetic—almost 40 percent of our population. These health issues are not incidental: They are symptoms of a food system that prioritizes volume, convenience, and profit over nutrition and well-being.

Looking ahead, the challenge becomes even more daunting. By 2050, the global population is projected to reach 9.7 billion, requiring 60 to 70 percent more food than was produced in 2010. And while we often speak of abundance, at any given moment, the world has only enough food in circulation to sustain life for about three to six months.

These facts paint a sobering picture. I'll admit this to you: I have sleepless nights over this stuff. The current food system is not just unsustainable—it is actively driving ecological breakdown and failing to meet the needs of human health. But I'm not raining down the bad news to depress you. Rather, I hope it's the opposite. Change is not a matter of preference or ideology. It is a matter of survival. The urgency is real. And the opportunity to do so is still in front of us.

# Here and Now

The message I tried to share with my Last Supper events, in retrospect, carried one crucial flaw. The warnings it sounded were all about events that would occur in a future, warmer world. It implied that if humanity has the foresight to curtail global warming in a timely manner, we will still be able to enjoy our familiar chocolate, coffee, and fine wine. If we don't, we will someday lose these pleasures—with *someday* being the operative concept.

Well, welcome to "someday."

In regions around the world, we are already too late to avoid the crises set in motion by changing climate. Food producers, whether they haul up empty crab traps off Oregon, watch as corn plantings sink beneath floodwaters in the fields of Nebraska, or try to feed a family from a tiny rice paddy in Indonesia, are foot soldiers in a rearguard action against an unrelenting, hostile climate. There is no one consistent effect of climate change. While the climate is warming, it's not as simple as the temperature going up each day. What a warming climate does on a macro scale is to produce far more extreme, more unpredictable weather events: more devastating storms; more drenching floods; longer and hotter droughts; migrating invasive pests and diseases. All of these are existential threats for whole seasons, or years, of crops, not to mention human livelihoods and life. On many fronts, climate is winning, providing a real-life preview of what a more uncertain climate has in store for human society.

Look, I'm not here to make you feel afraid and hopeless. In fact, anything but. Because I believe—and scientists, farmers, and technologists know—there are still ways of getting us out of the worst-case scenario. But we also have to know what is happening right now.

## FEELING THE PINCH

Even after three decades of fishing commercially in the waters off Oregon and Washington, Zed Blue had never seen a creature quite like the bird paddling along with a flock of gulls behind his boat in 2019. "It was huge," he told me. "A prehistoric-looking thing."

Blue, a strapping man in his forties whose brush cut caps a bushy red beard, snapped a picture of the bird with his smartphone and posted it on a nature website. Experts identified it as a northern giant petrel. Despite its name, the species is not in the least bit northern. It is native to the waters around Antarctica. It turned out that Blue was the first person to report seeing one this side of the equator.

The petrel's presence was one of many unusual changes Blue had noticed recently. Earlier, he scooped an exhausted brown-footed booby out of the waters of Puget Sound and turned it over to a bird sanctuary for rehabilitation. A normal enough humanitarian gesture, except that the species' normal habitat extends no farther north than Mexico. A colleague of Blue's hauled in his nets to find them full of silvery-blue albacore, native to temperate and tropical waters. He had never encountered them before, and had to radio authorities to inquire about the regulations for catching and selling the small tuna species. "What is going on that's causing these creatures—that have

evolved over tens or hundreds of thousands of years in one habitat—
to be coming all the way up to the far end of the planet from where
they live?" asked Blue. "It's alarming."

At the same time, the opposite is happening: Some familiar north-
western Pacific species were disappearing. Dana Wilson, a soft-spoken,
silver-haired member of the Makah tribe, whose homeland is on the
western tip of Washington's Olympic Peninsula, has fished near there
for salmon for most of his sixty-plus years, as had his father and
grandfather. He is worried that the family tradition will end. Salmon
catches had declined steadily until 2019, when there were no longer
any salmon to catch, period.

Dungeness crab fishing, worth $200 million a year, making it the
West Coast's single most valuable fishery, was once a major source of
income for Blue. He even served as vice president of the Washington
Dungeness Crab Fishermen's Association. Blue recalls that the sandy
bottom around Destruction Island, a few miles off the Washington
coast, was prime crabbing territory until the mid-2010s. "I caught an
insane amount of crabs there," said Blue. Today, there are almost no
crabs in the area, and the few that find their way into his traps are
limp and almost dead when taken out.

Elsewhere, places where Blue could fish for crabs and make good
money for several months became productive for less than two weeks.
The season, which starts each year only after fishery managers determine that the crabs have packed on a set amount of meat during
the warmer months, keeps getting pushed back later and later as the
crabs grow more slowly. Washington-based boats used to be able to
get out in December. Now the crabs don't fill out until January or
February, meaning crabbers have to endure fierce storms, bitter cold,
and monstrous waves during the most inclement time of year. Many
fishermen aren't making enough money to pay off their preseason
expenses. "I'm just a fisherman," Blue said, with a note of sad irony.

"Ain't got no book learnin'. I'm just trying to figure out the pieces. Why is this happening?"

NO ONE WOULD ACCUSE Nina Bednaršek, PhD, of lacking book learning. Bednaršek, who has clipped brown hair and speaks with an accent that gives away her upbringing in Slovenia, is a professor at Oregon State University, and was previously a senior scientist at the Southern California Coastal Water Research Project, an institution overseen by a consortium of public water agencies in the Greater Los Angeles/San Diego area. One of her areas of specialization is larval Dungeness crabs. Her work provides possible, and troubling, answers to Blue's question.

Laboratory experiments had shown her that—in theory at least—increased water acidity has adversely affected crab larvae, which are born about the size of grains of rice and drift on currents in the water column before settling to the bottom as juveniles. As acid levels increased, the crabs' shells became wrinkled and pitted, and hairlike sensory structures on the surface (called setae) fell out. Experimental animals raised in acidic conditions were smaller than their counterparts reared in buffered water.

Bednaršek originally viewed her lab experiments as purely theoretical and predictive, mimicking water conditions that might affect wild crabs after three or four more decades of increasing ocean acidification, which is driven by atmospheric carbon emissions; as that carbon is absorbed by seawater, it's converted into carbonic acid. She realized her error after examining Dungeness larvae captured during a 2016 field trip in the open ocean. The crabs were already displaying significant carapace dissolution. "We were shocked," she said. "It was no longer a prediction or a future scenario. We had not expected that degree of damage so soon."

Losing setae can decrease the survival rate of larval crabs. The sensors are essential for the larvae to find the right habitat, capture food, and avoid predators, particularly during the vulnerable period when they molt and begin metamorphosing into juveniles. According to Bednaršek, losing sensory organs can also reduce survival by changing swimming behaviors and limiting the crabs' ability to regulate buoyancy and maintain a vertical position. Small decreases in larval survival can have a large impact on what's available for fishermen like Blue to catch four or five years down the line, when the crabs have grown large enough to harvest.

"Conditions are changing, and they are changing rapidly," Bednaršek said. "And it's very likely that it will start happening on a larger scale sooner than we would have predicted. You can no longer be ignorant and think this will take decades. We don't have a decade. We have to develop management strategies now."

Prior to Bednaršek's discoveries, scientists had mistakenly assumed that decapods (crabs, shrimps, lobsters, and their kin) were less sensitive to ocean acidification than mollusks such as clams, oysters, and mussels. The two groups of organisms use different ocean chemicals to build their shells. But at a 2019 gathering of dozens of marine biologists from around the world that Bednaršek organized to try to establish thresholds where levels of ocean acidity would produce negative effects on Dungeness crabs and other decapods, researchers reported that the water was poised to cross these thresholds in several hot spots. "It's not just Dungeness crabs," said Richard Childers, who served as Washington state's ocean acidification lead. "It's California lobsters, it's Maine lobsters, it's shrimp." Sadly he was right. Indeed, in Alaska, officials closed the snow crab fishery for the first time in its history for the 2022–23 season. From 2018 to 2022, the snow crab population declined by about 10 billion crabs, going from 11.7 to 1.9 billion crabs, an almost 85 percent decline in less than five years. An ocean heat wave and acidification are what scientists blame as the culprits. Hopes of a rebound so far have not materialized, and in 2023–24 the fishery was closed again for the second straight season.

There may also be a connection between Zed Blue's empty crab traps and Dana Wilson's empty salmon nets. Although Dungeness crabs have few predators once they become juveniles and settle on the bottom, during their drifting larval months they are the primary food source for salmon, according to Childers. It's likely that one of the reasons for the decline is that the salmon can no longer find enough sustenance in some areas off Washington state.

## SHELL GAMES

You could tell that something bad was happening in the waters off the West Coast in the 2010s, when the farmed oyster industry in Oregon and Washington suffered near-total mortality among the young "seed" oysters raised in hatcheries.

I met Bill Dewey, then the director of public affairs for Taylor Shellfish Farms, in a former oyster-shucking room that had been remodeled into a restaurant at the company's facilities south of Bellingham, Washington. Gazing over Bellingham Bay, with its shaggy, conifer-covered islands, we snacked on some of the finest seafood I've ever eaten: delicious mahogany clams and an assortment of Kumamoto, Shigoku, Olympia, and Pacific petite oysters. It became disconcerting when Dewey started to describe the challenges of raising the luscious bivalves in increasingly acidic ocean water.

A tall, raw-boned man in his sixties, Dewey looked more like a rugged fisherman than a corporate executive (he did, in fact, operate his own small shellfish farm as a side hustle to his day job at Taylor). He explained that the company once operated its hatcheries without facing many problems, other than the occasional outbreaks of disease that they could clean up with medication and then proceed as normal, until the oysters grew to about the size of a fingernail, ready to be moved to farms in open water.

But toward the end of the 2010s, Taylor and other oyster farms began experiencing more failures than successes in their hatcheries. No one understood what was happening. At first, Dewey and his colleagues suspected bacterial infections. But that proved wrong. Finally, at a National Oceanic and Atmospheric Administration conference, a scientist gave a talk in which he reported that the culprit was ocean acidification. High acid levels were preventing the larval oysters from building shells.

"People were walking around with their chins on the floor, like a relative just died," Dewey told me.

I (and all of us, for that matter) wouldn't have been enjoying those Kumamotos—or any other West Coast oysters—were it not for the invention of a device that monitors water in hatcheries in real time and, before acidity levels become too high, automatically adds chemicals that buffer the acid's effects.

That invention bought the industry time, but oyster farmers are by no means rejoicing. They know that worse is on the way. Because of currents in the Northwest, coastal areas are fed by water being cycled up from extreme depths. Given the time it takes to complete the cycle, the water that is near the surface today last came to the top (the only place where it can absorb $CO_2$) between thirty and fifty years ago, when atmospheric levels of the greenhouse gas were significantly lower than they are now. What that means is that for the next several decades the oyster industry's fate here is sealed. Water with even higher levels of acidity will keep surfacing for decades, even if we curtail $CO_2$ emissions today.

Once seed oysters are big enough to be moved to the open water, they can tolerate current acid levels. "But it's only a matter of time before the acidity is going to start having an effect on juveniles," said Dewey.

In open water oyster beds, farmers will not be able to add buffering chemicals, as they do in enclosed hatcheries. "Maybe we'll be able to grow oysters in seaweed beds, which naturally offset acidity," said Dewey. "Maybe we could breed resistant oysters, or use smaller species that are naturally resistant—if consumers will accept them. We are committed to making it work. We're resilient. We won't go away. But one thing for sure, if we don't start dealing with the source— carbon in the atmosphere—the future is going to be grim for a lot more than shellfish."

Zed Blue learned all about what that grim future looks like. After a difficult year in 2020, the bank foreclosed on his boats. When we met at the end of that year, he told me he wanted to stay involved in the industry but had yet to figure out how.

## AGAINST THE GRAIN

The family farm that John Nowatzki grew up on in the 1950s and 1960s in North Dakota is unrecognizable today. For three generations, his family, like most others in the area, cultivated wheat, along with a little barley. No one imagined being able to successfully grow corn on the nine-hundred-acre spread, which is only four miles from the Canadian border. During his boyhood, Nowatzki, who left the farm to become a professor in the agriculture department at North Dakota State University, had never seen corn growing anywhere near the homeplace. Until recently, the warmth-loving crop was unknown north of Interstate 94, which traverses the state nearly two hundred miles south of the Nowatzki farm. North of I-94, the summers were too cold for corn, and the growing season too brief. Today, however, the Nowatzki fields are covered in corn and soybeans. Wheat grows in only a tiny sliver of the property.

What accounts for the shifts? Nowatzki, a spry man in his mid-seventies with a shock of pure white hair, said newly developed shorter-season corn varieties have contributed, but the big reason is the warming climate. His father never planted varieties that required more than ninety-five days between the last killing frost of spring and the first of fall. The risk that the crops would be destroyed by cold was simply too great. Now the growing season is more than twenty days longer than it was in his father's era—plenty of time for warm-weather crops to mature. Warmer weather also allows the air to hold more moisture, which brings the formerly dry area adequate summer rainfall for corn—a thirsty crop.

Corn first started to appear in northern North Dakota in the early 2000s. In the intervening years, the state's acreage of the crop has quadrupled. And for the time being, growers there are loving it. Traditionally, corn is a much more profitable crop than wheat.

But what climate change giveth with one hand, it taketh with the other. The weather in the northern plains has become increasingly erratic. In 2019, a late, wet spring prevented farmers from getting an early start with their planting. A cool, rainy summer that slowed corn's growth followed. The state then experienced an early, heavy snowfall in October. The snow prevented farmers from getting onto the fields to harvest. Fully half the corn crop was left standing through the winter. It was the second time that had happened in recent years.

Current heavier rainfall patterns also allow farmers to grow corn in parts of North Dakota's fertile Red River valley where small grains, which require less water, were once the dominant crops. But it's become a case of too much of a good thing. Salt-laden bedrock underlies much of the valley's soil, and a rising water table is bringing that salt closer to the surface. Tom Sauer, who is with the National

Laboratory for Agriculture and the Environment, said that his brother-in-law made good money raising corn in the valley for several years—until his land became so salty that he could hardly grow a crop, period. Corn's growth habits exacerbate the problem. Wheat, barley, and other small grains are shallow rooted and absorb most of the water they need early in the season. Corn, by contrast, is deep rooted and sucks up the most moisture late in the season, drawing salt to the surface. The U.S. Department of Agriculture (USDA) predicts that parts of the Red River valley will be too salty to produce crops in thirty years.

Although North Dakota farmers on balance still benefit from climate change, their peers to the south—the heart of the nation's corn belt—are feeling global warming's downsides. Fields in the Missouri River valley in Iowa and Nebraska were flooded for much of the spring and early summers of 2019 and 2020. The problem has become so acute that there is talk in government circles of paying farmers to take the land out of crops and allow it to revert to wetland and wildlife habitat. A "250-year flood" is an event the scale of which has historically happened only once every 250 years. Ames, in central Iowa, has experienced *three* 250-year floods in the past twenty-five years.

Conversely, in 2012, crops withered in the same area from a severe drought. "A lot of these climatology records have to be rethought," said Sauer. "Farmers are not interested in debating who or what is causing these changes. They are trying to deal with the reality they have."

## AG SCHOOL

Anticipating his home state's quadrennial invasion of presidential hopefuls in 2019, Eugene Takle, then the director of the Iowa State

University Climate Science Program, prepared a simple, single-page handout on climate change and how it affects agriculture in the nation's premier corn-growing state. One page long, he said, because he wanted it to be of a length that politicians would read. The front side listed current effects; the flip side future ones. To make it easier for the politicians to understand, he color-coded the document, green for beneficial changes, red for bad. "You didn't even have to know how to read to comprehend the document," he said wryly.

The side listing present effects was about half green and half red. But the reverse, reflecting future trends, was entirely red. On the positive side, farmers could plant earlier. The growing season was longer. Rain fell late in the season when corn needs it most. And there were fewer extreme heat waves. (Excessive warmth prevents corn from developing kernels.) Negative effects included warmer soils that would allow pests to overwinter. Increased humidity in the spring would further benefit those pests. Greater concentrations of soil moisture would create costly drainage problems and delay planting. To compensate, farmers would need bigger, more expensive machinery to seed a crop that would not be worth any more money. "At some point, they are going to say 'Enough!'" Takle said.

Fully one-quarter of corn's yield increases in recent decades is attributable to favorable warmer weather, kept stable by cooling effects when rainfall evaporates. But when excessive heat begins to reduce rainfall—as it will—the underlying global warming that has been modulated by evaporation will make itself manifest. In southern areas of the corn belt, yields are already dropping. With higher temperatures, even irrigation will not save corn. The warmest day of the year in Iowa is expected to rise by seven degrees in the near future. Heat waves will be thirteen degrees hotter and more frequent than now. "You won't be able to pump enough water to suppress the damage from that amount of heat," Takle said. "Water can't move fast enough through the plant to cool it. Some drier areas are going to have to look for other crops."

Even in rainier areas, he expects corn yields to drop by 25 percent by 2050 as floods alternate with droughts to plunge even more farmers into financial ruin. All of the agricultural productivity gains made between 1980 and 2010 will be reversed. The title of one of Takle's recent papers in *Physics Today* is "Iowa Agriculture Is Losing Its Goldilocks Climate."

The same could be said for many of the rest of the world's most fecund regions.

## HOT COFFEE

Some of the finest coffee anywhere comes from the small town of Santa Maria de Dota, nestled in a bowl-shaped compression in the south-central highlands of Costa Rica. Jungle-cloaked peaks rising to seven thousand feet above sea level encircle the community, with its symmetrical grid of streets, red-roofed houses, and adobe-colored church steeples. The hillsides that surround the town are so precipitous that it's difficult to see how any plant could take root there, but a close look reveals that the inclines are dotted with ranks of small, rounded coffee trees.

In the world of coffee, where most growers are peasants who barely eke out a living, Dota is a kind of Shangri-la. The local cooperative, Coopedota, processes and ships the coffee beans grown by eight hundred smallholder members. Thanks to high demand from specialty roasters in Europe and the United States, Coopedota beans can fetch prices 40 percent higher than the going market rates.

The cooperative invests some of its proceeds in infrastructure projects and social programs that benefit all town residents. Its quiet, litter-free streets (the co-op pays for garbage removal and recycling) are lined with boutiques and mom-and-pop cafés painted in pastel blues, yellows, and pinks. Groups of children on their way to school wear white dress shirts, neckties (both boys and girls), and pressed navy-blue trousers. The kids take classes that prepare them for good jobs in the area after graduation. They enjoy a wide selection of sports programs. Dota's senior citizens can pass their retirement in a facility that bears more resemblance to a chic eco-lodge than a typical drab American retirement facility. A store sells farmers agricultural supplies at discount prices and will advance them funds to be paid back when the crop ripens. Coopedota pays for all of this.

As the director of the co-op for a quarter of a century until he retired to focus on tending his family's small coffee farm, Roberto Mata Naranjo deserves credit for much of the prosperity his town enjoys. He initiated the policy of cutting out profit-hungry, often corrupt middlemen by dealing directly with select foreign roasters who understood the value of paying premium prices for Dota coffee. But these days, he wonders how long the good times will last.

Formerly, Dota farmers grew coffee beans at elevations of between five thousand and fifty-five hundred feet above sea level, where temperatures hovered between sixty-five and seventy degrees, the sweet spot for the finicky bushes of the *arabica* species that produce the majority of coffee that we drink, because it's considered the best species of coffee. The *robusta* coffee species, a cousin to *arabica,* is much less susceptible to temperature variations, but its harsh flavor largely limits its use to being a base for instant coffee. Unfortunately, as the heat increased, Dota farmers were forced to move plantings to elevations of sixty-five hundred feet above sea level or more. If the trend continues, even the highest land in the area won't be temperate enough for arabica.

Of particular concern to Mata, who likes to use a minimum amount of agricultural chemicals for both environmental and economic reasons, warmer weather has also allowed a fatal fungal disease called coffee rust to invade Costa Rica and most of the other arabica-growing areas. The leaves of infected bushes first show the reddish-orange spots that give the disease its name. They soon yellow, then blacken, curl up, and fall to the ground. Eventually the trees die.

Coffee rust was first identified in Ceylon in the late nineteenth century, where it completely wiped out production in what was then the world's largest growing region. The fungus was contained in the Eastern Hemisphere until 2008, when outbreaks occurred in Colombia. By 2013, warmer temperatures allowed it to journey north into Central America. In those days, treatment with fungicide would take care of the disease, albeit at considerable cost. But in recent years farmers noticed that the chemicals that formerly stopped rust outbreaks no longer worked.

At first horticulturists thought that rust, like many disease organisms, had developed resistance to fungicides. They were wrong. The culprit was something much more difficult to deal with: heat. With warmer temperatures and the increased rain that they bring, rust can reproduce and spread before the chemicals destroy it.

One problem that leaves arabica particularly susceptible to rust and other diseases is that the plants alive today lack genetic diversity. They are a sickly, inbred lot, not unlike some lines of purebred dogs and royal families of medieval times. Prior to the late seventeenth century, Yemen kept strict control of coffee cultivation, carefully guarding plantations to prevent foreigners from sneaking in and purloining plants to be grown elsewhere. But toward the end of the seventeenth century, traders smuggled a few trees out of the country and planted their offspring throughout the so-called coffee belt, which girds the world between the Tropic of Cancer and the Tropic of Capricorn—East Africa, Central and South America, Hawaii, and Southeast Asia. That handful of purloined plants became the genetic foundation for every arabica tree grown today. This paucity of genetic diversity means that arabica lacks the gene traits that allow its wild relatives to fight diseases and pests—necessary resilience for survival in a changing climate.

A hotter planet has distressing implications for coffee wherever it is grown. South African researchers found that when nighttime temperatures rise by slightly less than two degrees, yields drop by 40 percent. They conducted their study in Tanzania because the climate in its highlands is representative of that in most other coffee-growing areas, including Brazil, Colombia, Kenya, and Costa Rica. Researchers predict that global warming will shrink the prime growing area for arabica beans by half over the next twenty years. This translates to reduced production and substantially higher prices over time.

Plant breeders are racing to develop strains of coffee that resist rust. Two obstacles stand in their way. One is that creating new coffee varieties takes decades, time that growers like Mata don't have. The other is that just when breeders need all the genetic help they can get, the wild coffee species that researchers look to for resistant genes that they can breed into domestic plants are also in steep decline due to climate change. Botanists find themselves in a race against time to find as many as they can of these critical wild species before they go extinct. Scientists at England's Kew Gardens speculate that by the end of this century 85 percent of the areas where wild coffee has traditionally grown will no longer support it, causing 80 percent of wild species to disappear. The Kew team adds that two-thirds of all coffee species, including the ones we drink, are at risk of going extinct.

In an interview with *The Economist,* Vern Long of World Coffee Research, an industry-funded group, had a stern warning for everyone who regularly enjoys starting each day with a cup of coffee. "If we don't have the innovation to respond to climate challenges, we're just going to be drinking synthetic coffee."

Gulp.

## IT'S EVERYWHERE

I chose to focus on Dungeness crabs, oysters, corn, and coffee in this chapter because they provide stark examples of what climate is doing to food production—*now.* But I could have gone to almost any crop-growing region in the world and found farmers engaged in similar struggles with erratic weather.

The American state of Georgia may someday have to search for a new nickname. Warming winter temperatures mean that the Peach State's iconic fruit trees no longer get the consistently cold weather, or "chill time," required to produce robust buds, resulting in smaller harvests of poorer-quality fruit. Occasionally the lack of chill time yields financially devastating crop failures.

Warm spells during the spring in Wisconsin have caused the state's apple crop to bloom too early, only to have blossoms destroyed by frosts. Blueberries in Maine and sour cherries in Michigan have suffered similar fates. Mild winters in New York state have provided ideal conditions for the fruit flies that feed on organic raspberries.

Castroville in central California bills itself as the artichoke capital of the world. Unfortunately, the moist, cooling winds wafting over the fields from the Pacific that the artichoke varieties grown around Castroville need are becoming less reliable, forcing farmers to turn to new varieties suitable to the desert heat in the southern part of the state. Avocados in Mexico, olives in parts of Italy, pistachios in Iran—the list of climate-affected crops quickly becomes mind-numbing.

If I had to nominate a crop that best exemplifies the damaging effects of atmospheric $CO_2$ on food production—and how producing food itself leads to the creation of climatic conditions that make farming more difficult—I would vote for rice. While Americans and Europeans might view rice as an innocuous filler for beef, lamb, and chicken recipes, it has been *the* staple of Asian cuisine—and therefore of most of the world's population—for eons. Today, Asians consume an average of 170 pounds of rice per person per year. Nearly 3.5 billion people depend on rice for their survival. Asia grows nearly 90 percent of the world's rice. Yet after years of increases during the so-called Green Revolution (when improved crop varieties and new technologies and chemicals allowed for much more efficient, if less sustainable, agriculture), rice yields are tumbling, even as demand continues to surge.

Climate change deserves much of the blame for a host of challenges facing rice farmers across the continent. Higher temperatures cause the notoriously thirsty plants to shrivel in the caked earth of dried paddies. For each rise in minimum temperature of one degree Celsius, rice yields drop by 10 percent. Floods in Pakistan in 2022 destroyed 15 percent of its rice crop. Higher sea levels have flooded low-lying coastal paddies in areas such as Vietnam's Mekong delta with salt water.

Rice also actively contributes to its own difficulties. Submerging soil underwater in paddies, which deprives it of oxygen, allows populations of methane-producing bacteria to explode, pushing tons of the potent greenhouse gas into the atmosphere, enough to make rice agriculture's contribution to warming emissions equal to that of the entire airline industry.

## PREPARING FOR DISASTER

Exactly how urgently action needs to be taken became clear when I sat down with Sara Menker, who ran Gro Intelligence, a company she founded in 2014 with backing from prominent Silicon Valley venture capitalist groups. Gro has since closed its doors, but to this day, it was the most sophisticated data and analytics company working on the intersection of agriculture and climate change that I have known. Menker applied advances in artificial intelligence to forecast trends in food production in an era of climate change. "It's about getting ready for disaster," Menker described her company's work to *Time* magazine in 2021. "It's about hedging for downside risk."

To accomplish that, Menker and her team analyzed fifty thousand data sets and had more than two million models in its computers—crop reports, weather conditions, agricultural labor issues, soil moisture levels, livestock diseases, insect infestation—and produced an incredibly sophisticated set of insights into what we are facing.

At a time when most experts accepted the World Bank's 2009 estimate that humans will have to produce 70 percent more food by 2050 to adequately nourish the world population—a daunting task, but far enough in the future to permit a certain complacency—Menker's newer, much more sophisticated data analysis showed that unless dramatic steps are taken immediately, the crucial tipping point beyond which the world will no longer be able to feed itself will arrive before 2030. It's vital to note that many of the most horrible episodes of famine and mass starvation in history have been the result of political or logistical issues; almost always those catastrophes came because humans lacked the ability or will to get food that actually exists somewhere in the world to hungry people. Indeed, we as a planet continue to produce far more food than people eat, and food waste is a massive issue on its own. But the incredibly dark future

Menker described is a world that *cannot physically produce* enough food to feed humanity. "The train wreck is right in front of us," she told me.

Menker, who was recognized as a Young Global Leader by the World Economic Forum, learned about food shortages firsthand in the 1980s during her childhood in Ethiopia. Although her immediate family was comfortably middle class, always had food on the table, and could afford paying her tuition to attend a private school, millions of her compatriots suffered from dire famine and the violence sparked in no small part by food scarcity. She earned her undergraduate degree at Mount Holyoke College in Massachusetts and went on to do graduate work at the London School of Economics and Columbia University, before becoming a commodities trader at Morgan Stanley. Propelled by a fierce intelligence and a superhuman ability to translate complex computer calculations into hard information businesses could act upon, she rose quickly to the rank of vice president.

Although her specialty at the investment bank was energy futures, her passion—instilled by the uncertainties she had observed as a child in Africa—was the future of agriculture. During the 2008 recession, she recognized that there was a potential for a global food disaster and decided to direct her analytical talents toward finding ways to cope. Her personal obsession became a profession when she gave up her career and launched Gro with the goal of disrupting traditional approaches to predicting worldwide agricultural output.

Traditionally, agricultural estimates were based on bulk measurements—so many bushels of grain, so many pounds of beef. Menker saw this as fundamentally flawed. After all, not all foods are created equal. A pound of pork is not equal in nutrition or calories to a pound of flour, which is not equal to a pound of sugar, facts she learned the hard way when she gained weight after coming to the United States, even though she ate the same quantity of food as she did in Ethiopia. The difference was that her new diet was more nutrition dense and packed more calories.

Menker decided to take a different approach and measure food production in calories. The numbers she produced were clear and frightening: By 2030—only a few years from now—the world will be producing 214 trillion fewer calories per year than what's required. For scale, that's about the number of calories consumed by the entire population of the United States.

Menker's projections showed that China, which does not grow enough to keep its citizens fed at present, will continue to have to import vast amounts of food to meet the demands of its increasingly wealthy population. Meanwhile, in Africa, where agricultural production has languished, the population will expand to surpass China's. Those new mouths will have to be fed somehow. And India, which has been able to feed its more than one billion residents in recent decades thanks to innovation, will flip. Like China and Africa, the subcontinent will become a net food importer. Taken together, this means that half of the world's population will be looking beyond the borders of their home countries to secure enough to eat. Millions won't be able to find it.

Yet, even in the face of such realities, Menker remains cautiously optimistic. "We have the solutions," she told me. "But this is probably our last chance to get it right. We have to act."

What keeps me from becoming overwhelmed by the dismal predictions was the resiliency of producers I encountered and their willingness to change. Zed Blue might have lost his boats to the bank, but he is determined to remain a fisherman. Bill Dewey at Taylor Shellfish takes comfort knowing that his industry has had to adapt to oyster epidemics in the past, and, he thinks, it will adapt in the future. "Maybe we'll be raising something different—who knows? Jellyfish?"

he told me. I couldn't tell whether he was joking. Coffee growers like Mata have begun to plant new varieties, hoping to find ones that survive rust and produce adequate yields of quality beans.

Producers need all the support they can get to bring about the necessary changes in time to stave off full-blown food disasters and meet the challenges of a warmer, more volatile world. My time at the White House showed me that making the necessary changes in the food system was possible—but difficult. To do so, we must address the four pillars of change: culture, government policy, business, and technology.

# PART II

## THE FOUR PILLARS
## OF CHANGE

# Changing Culture

Humans are naturally drawn to neat, compelling narratives that simplify complex issues into clear-cut battles of good versus evil. It's a storytelling instinct that helps us make sense of the world, but it also leads us to oversimplify the real problems we face. When it comes to the state of our food systems, the dominant story points the finger at big business and flawed government policies, positioning them as the evil villains who exploit consumers and degrade the planet. The story we have been fed—if you're reading this, I assume you're in the camp that has heard this—is that government subsidies are the reason farmers grow what they grow. These subsidies are a result of collusion between big business and the government; they reduce the price of unhealthy food and are the driving force of why we eat what we eat. The story continues to say that even though the people are demanding healthy, regeneratively produced food, the government is unwilling to act and companies are unwilling to produce it. There are some elements of truth to this narrative, and the idea that big business has an ability to influence policy is certainly true. But government ignoring the will of the people to produce better food in better ways is not the reason our politics and our businesses are failing us. While this tale is easy to digest, it overlooks a more uncomfortable truth: Voters and consumers, the very people these systems serve, are complicit in the problem. At the root of this is our culture.

The reality is that most people don't prioritize climate change or regenerative agriculture when making decisions about food and politicians. Most don't even consider it. Convenience, price, and taste reign supreme in consumer choices, and as a result the market responds to those demands rather than focusing on sustainability. Right now, in the mass market, no one cares about regenerative agriculture (or has any idea what it is), and people are simply not willing to pay more for sustainably produced food. There are a number of reasons for that which we will discuss later, but this is a massive barrier to change. Similarly, politicians tend to craft policies that align with voter priorities. Right now, voters are not voting on climate policy, let alone food policy. In essence, while corporations and government are failing us, they are simply not going to change unless we change. Both our businesses and our politicians are reflections of our cultural values. We have not yet overcome the most fundamental obstacle to progress: widespread apathy or disinterest among the people who have the power to drive meaningful change. Our businesses will keep degrading the planet if consumers don't purchase differently. Our politics will continue to fail if voters do not vote on these issues.

This is starkly reflected in the data around voter priorities in the 2024 election. According to a Pew Research Center survey, only 38 percent of registered voters consider climate change a "very important" issue in deciding their vote, placing it well behind concerns like the economy (81 percent) and health care (65 percent). Even among younger, more environmentally conscious voters, climate change often ranks behind issues like student debt and health-care access. In fact, a Gallup poll found that while 66 percent of Americans recognize climate change as a major issue, only 11 percent view it as their top concern for the election. And that is climate change as a whole; for the vast majority of voters, food policy is simply not a top issue.

These numbers reveal a sobering truth: While climate change is widely acknowledged, it is not treated with the urgency it deserves in the political arena. Voter priorities shape policy, and if the electorate doesn't demand action on climate and sustainable food systems, politicians have little incentive and mandate to act, even if they actually care about the issue. But mostly, the reality is worse: The people we send to office tend to reflect our values, and the majority of us are simply not invested in these critical issues. Of course, democracy is *not* simply about voting and winning massive majorities at the ballot box: It's also about calling your elected officials, going to town halls, and showing them that this issue does matter. The real challenge, then, is not just holding businesses and governments accountable; it's that to do so, we need to be changing the mindset of the mass of voters and consumers who underpin these systems.

There is no getting around it. I cannot emphasize that enough. The hard truth is that we have to shift the core values of Americans' food culture before we can bring about the changes in the food system we so desperately need. We may land some wins here and there, but the cultural barriers we currently face are a massive roadblock to true systemic change. Doing so is the hardest and most important thing to accomplish if we want to see food produced in a manner that harms neither the environment nor the humans who eat, and we need to do so in rapidly changing climatic conditions.

True, thanks to advocates' efforts, "organic," "local," "seasonal," and "artisanal" became buzzwords, and food culture *has* changed, but only for a minority of the population—primarily conscientious eaters who live in a bubble of distorted culinary reality. I am one of them, as are nearly all of the people in my social circle. I frequent "locavore" restaurants. My garden is the source of most of the produce my family eats in the warm seasons, with much of the remainder coming from nearby farms. But in reality, even after a couple of decades of exposure to the so-called food movement's mantra, the vast majority of Americans give little thought to the food they eat aside from its cost, convenience, and taste. I don't mean that as a criticism; it's just seeing reality and understanding how much culture-change work lies ahead of us. Sales of organic food have doubled since the early '00s. Which is great. But organic still represents a mere 6 percent of total grocery sales. Farmers markets, those cornucopias of local fare, have more than quadrupled in number since the turn of this century, but today local food represents a minuscule 1.5 percent of agricultural production's value. So far, good-food advocates' efforts, however praiseworthy, are not nearly enough to get the food system to where it needs to go in time to stave off catastrophe.

Although well meaning, the embryonic food movement preached to a choir of like-minded souls, but we neglected to consider how government and business leaders actually make decisions that have the potential to create massive changes in food production. As I learned in Washington, these leaders can safely ignore the entreaties of a well-meaning few, but they take careful measure of the majority of their constituents—voters for politicians, customers for business executives. The key question leaders ask before taking any action is, What changes do our constituents want? What do they need? Neither politicians nor business executives will take on much risk unless they are pushed by constituents and stakeholders, and believe the risk will pay off. Only after they are confident the conditions will result in positive outcomes do legislators and managers of successful businesses begin to take action in earnest. Pressure to act is important, but in the end a smaller part of the equation. Winning on the other side of risk taking will drive far more change. Meaning, if a politician successfully moves us in the right direction, do they see fundraising numbers go up and improved chances of winning the next election? If the answer is yes, we will see more and more action on the issues we care about. If a business leader launches a new regeneratively sourced product, do they see sales spike and continue to grow? If the answer is yes, we will see more and more products launching with sustainable ingredients. Our cultural values underpin our politics and business.

Unfortunately, shifting people's attitudes toward food is an amorphous task. Progress, if we are fortunate enough to get it, is painstakingly slow and can be boring—the opposite of a silver bullet. It often comes, if it comes at all, in imperceptible increments. Changing culture is also messy and complicated. This is true in any area, but doubly so with food culture. In tinkering with it, you are messing with people's very identity. Food is one of the deepest expressions of our identity. It is how we express who we are and who we are not. It is how we show love to ourselves, our families, and our communities. What we eat ties us to our ancestors and our childhoods. When we are told what we are eating is bad, the implication is that *we* are bad. No one wants to hear that from anyone else, let alone the government. And the tricky part is that everyone happens to be an expert on food. After all, most of us eat at least three times a day.

No wonder food activists generally haven't focused effort on the challenge of culture and instead move directly to try to change government policies and politicians' priorities. But that shortcut dramatically limits what we can accomplish through policy. It's no wonder why politicians are generally very cautious when approaching any major food-related policy issue. Politicians will never move forward unless they feel the pressure of culture pushing them in that direction. This failure to lay the necessary cultural groundwork is the main reason why we've seen so little in the way of concrete change we hope for.

From the very beginning, Michelle and I viewed shifting the nation's food values as the single most important thing we could do. The good news is that we, together with the populace at large, made some progress and learned a few skills on how to do just that.

## SODA WARS

I learned in the hardest possible, most direct way about the absolute necessity of addressing cultural norms before attempting to make change. Early in my White House tenure as the Obamas' senior nutrition policy adviser, I made a colossal blunder by trying to implement my health-forward agenda within the White House itself, without considering how the people affected would feel about it. We had just launched Let's Move!, Michelle's initiative to shift how children and families ate and put them on a path toward a healthy future. Aside from the bluster of the inevitable detractors determined to criticize anything accomplished by the Obama White House, Let's Move! received a near-universally positive initial reaction through projects like the White House Garden.

Michelle wanted us always to walk the walk, and tasked me to ensure that the meals we served in the White House and various events like the Easter Egg Roll reflected the values that we were espousing to the nation. Hoping to build on this success, or at least not undercut it, I became concerned about a bizarre perk enjoyed by the 450 or so White House staffers: free Coke. That's right. Anyone working there could drop by the serving counter at the Mess, the name of the employees' dining facility historically operated by the U.S. Navy (not to be confused with the formal kitchen where state dinners were prepared), and pick up a can, a couple of cans, or a whole carton if their hearts desired—unlimited quantities at absolutely no charge. Rumor had it that the custom went back to the

Carter administration, or maybe earlier. Whatever its genesis, by the time we got there, complimentary Coke had become part of the White House DNA.

Coke! The sugary symbol of our national obesity epidemic. Even a rudimentary understanding of the microscope under which the Obama administration operated in predatory Washington should tell you that this was a PR catastrophe waiting to happen. Imagine Sean Hannity leaping all over that gossipy tidbit. I could all but hear the ginned-up outrage. "Oh, look at the two-faced nanny state." "Here's the holier-than-thou First Lady on TV dancing with Ellen DeGeneres, telling all of us to exercise and eat healthy food, while her staff enjoys all the free Coke they can guzzle." "How dare she tell America's mothers to stop letting their kids have soda when she hands out gallons of the stuff to her own employees!"

I decided we needed to preempt the looming backlash. My plan was hardly radical. Let the White House workers enjoy all the Coke they liked, but have them pay for it, just as they had to pay for a bottle of water, which, by the way, we were encouraging consumers to substitute for sugary beverages at the time. It sounded perfectly reasonable because it was. So, I brought it up with Michelle at seven o'clock one morning when she, the president, and I were working out with Cornell, our trainer and friend, as we did together most days.

Joining the Obamas for exercise each morning was one of the many ways my personal relationship with the First Couple could understandably cause friction with other members of the White House staff, particularly those who were my peers or my superiors. Our morning workouts were bracketed at the end of the day when Barack and I relaxed most evenings after dinner with three games of pool before he headed into his office to pore over a tall stack of dense briefing documents until late into the night. The gym and the pool table were spaces where we could preserve a bit of normalcy, cracking jokes, talking trash, and discussing everything under the sun. Those times were probably the most memorable I spent in my White House years.

The Obamas and I had become friends during the simpler, early months of my employment at their household in Chicago. I was nothing more than a young personal chef hired to cook for the girls and Michelle, whom I had known of since my high school days in the close-knit Hyde Park community near the University of Chicago. At the time, Barack was a first-term, junior senator from Illinois with aspirations to become president, a prospect that seemed so far-fetched as to be unimaginable. Michelle had spent years juggling the demands of being a student, a lawyer, an advocate, a wife, a working mother of two, and the partner of an ambitious politician. Out of necessity, healthful, prepared-from-scratch dinners often got short shrift.

"Like many parents, I was shopping primarily for convenience and cost," she confessed to a group of food business executives shortly after our arrival in Washington. "I bought products that were pre-packaged, precut, precooked. If it was 'pre,' I was getting it because I was looking for anything that was quick and easy to prepare and consume. And I was grateful for the time and effort that I saved with these kinds of products. But I was also completely unaware that all that extra convenience sometimes made it just a little too easy for me to eat too much, for my kids to eat too much, and to eat too often. And like so many families, my family fell into the habit of living that 'grab-and-go' lifestyle, eating more and more between meals. And slowly, all those calories really just started to add up."

When the family's pediatrician told Michelle that her children were on the path to becoming overweight, she decided to make a change. My assignment as their personal chef was to replace all that caloric "pre" stuff with from-scratch dinners that emphasized fresh fruits and vegetables at the expense of sugar, fat, sodium, white flour, and big hunks of red meat. Not only did the kids love the new fare, they soon began thriving, according to a subsequent pediatrician's report. Michelle noticed that the entire family was enjoying better overall health dining on my meals.

With Barack away in Washington or out on the hustings, the future First Lady and I spent long evenings leaning over the kitchen counter talking about food and its effects on her kids.

These conversations created bonds that we carried into the White House, where our friendship and my role as a big-brother figure to the children put me in a unique position relative to the established office chain of command. Everyone I worked with in the East Wing knew that I could execute an end run at any time around official protocol and go directly to the boss. Needless to say, that created a complicated dynamic. For good reason, my co-workers saw my personal relationship with the Obamas as a threat, although I rarely used my access to the President and First Lady in ways they imagined I did. It made me something of a target. I learned that I needed to tread carefully.

## NO, CHEF!

My culinary background also caused problems. White House colleagues constantly tried to use my profession against me. White House staffers stuck me with the label "the Chef," even though I officially had a dual role: family chef *and* senior adviser to both Michelle and the president.

Attempting to counteract the image of a cook, I made it a point to change from a chef's white jacket to a business suit and a tie when in my adviser mode. As I walked down the East Wing hallway on my way to one of our initial meetings with Michelle to refine the plan for the launch of the White House Garden, one of my fellow senior staffers looked up, stared at me dumbstruck for a millisecond, and blurted, "What are *you* doing here, and why are you in a suit?" The message was that I had no place dressing like someone who actually made important policy decisions. I resisted pointing out that a mere technocrat had no business making decisions about food policy. But the "You're just a chef" vibe continued to dog me and was used both internally and externally to derail Let's Move! efforts throughout my time in Washington.

Because I thought ending free Coke would be no big deal, I freely shared my concerns with Michelle about the risks of bad publicity from the policy and the necessity to protect her from being made to look like a hypocrite and having the good vibes being generated by Let's Move! overshadowed by what was outwardly a silly, trivial custom. No one on the staff made a fortune, but it wasn't as if they couldn't afford to part with a buck or two for the occasional soft drink.

Michelle, who was adamant that she and her family live according to the values she was publicly espousing, agreed without any discussion. "Free soda doesn't make any sense," she said.

I took the request to the navy officer who ran the Mess. He shrugged and said it didn't matter to him. Being military, he believed that whatever the president and First Lady wanted was what they got.

Two hurdles cleared.

Confident that my idea was a no-brainer if there ever was one, I came to the next managers' meeting ready to present our new Coke policy as a fait accompli, neglecting to float trial balloons and asking for feedback from my co-workers. Michelle's chief of staff, Tina Tchen (my immediate superior), her deputy, Melissa Winter, and I gathered to update each other on routine stuff. When my turn to speak came, I casually mentioned that Michelle and I had talked things over and that we could no longer hand out free Coke. I was prepared to explain our thinking, but before I opened my mouth, Winter exploded.

"What are you talking about?" she demanded.

In our high-stress environment, Winter—now a good friend—and I had a fraught working relationship, so I proceeded cautiously. "We are mounting a huge health campaign," I said. "We shouldn't be giving out free Coke. That is a total contradiction! Think of the negative PR."

She snorted. "You aren't serious. You've got to be kidding me," she said, her voice quiet and steady, but having seen her in action, I knew that she was inwardly seething.

Trying to put on a positive spin, I said, "Staff will still be able to buy it. Just not get it for free."

"It's the only free thing we get around here. People work themselves to the bone here. They make little money; they sacrifice time away from their families. They are exhausted. The least people should be able to get is the occasional free Coke."

"But how can we run a health campaign if we're still handing out free Coke?"

"Sam, you are not going to take it away, period. No way. Over my dead body."

The meeting moved on to other topics, but Winter—Mel, to us—called me into her office afterward, closed the door, and immediately tore into me. Heated exchanges were a common mode of communication between Mel and me, but this one went off the charts. We parted ways, both defiantly holding our ground.

Being a political animal, like most successful denizens of the White House, Mel wasted no time coolly and efficiently going on the offensive. She obviously had legitimate concerns that ending the Coke policy would dampen employee morale, which I could understand. But her motives were not completely altruistic. Mel was also a confirmed Diet Coke–aholic. During workdays, I rarely saw her without a can of Diet Coke nearby. It was truly part of her identity. Soon, word of the free-Coke ban started getting around. Colleagues began accosting me in the hallways. At first, they seemed to be in a state of denial.

"I heard a rumor . . . It can't be true."

"You're not really going to do this, are you?"

As it became clear that we really meant to end complimentary Coke, disbelief gave way to anger.

"Just who do you think you are, taking away the one damn perk I have?"

"You're not going to take my Coke from me. No fucking way."

Only half jokingly, I mused that death threats were going to be coming my way next.

Mel made sure the firestorm spread to the West Wing, where someone gave the most powerful man in the world an earful about the festering cola crisis. Barack passed the message along to Michelle. He obviously had a lot more important things he was dealing with and didn't need the team up in arms. It became obvious that my little no-brainer had erupted into a full-blown employee relations shit-storm. Michelle and I reconvened and decided to retreat before more damage was done. This was not the hill we wanted our efforts toward improving the nation's nutrition to die on. As quickly as she'd okayed the plan, she said that we had to let the office know that we had reconsidered our decision. I immediately rescinded my fatwa on free Coke.

Without doubt, the First Lady had the power to prevail, but at the cost of alienating the very people we had to closely work with to implement our broader goals. I certainly could have leaked what was going on to the press, and free Coke would have been abolished before the next news cycle had run its course. But that, too, would have been counterproductive.

Even though the debacle took place in the tight, insular atmosphere of the White House among like-minded colleagues, I learned a lot that I could apply to making change in the broader culture.

Thinking about what had gone wrong, I realized I had made several fundamental mistakes. As silly as it seems, those Cokes represented a treasured ritual in the White House. Everyone there worked under extraordinary pressure. The hours were interminable; the salaries far lower than what the private sector offered for similar jobs. Going down to the Mess at 10:30 in the morning and grabbing a Coke made even the most haggard worker feel a little bit special. Staffers could do something nice for themselves as they tried to deal with the insanity of 1600 Pennsylvania Avenue. I was so fixated on the liquid in those cans that I'd failed to consider their emotional contents.

I took the lessons of the fiasco to heart. No one reacts well to arbitrary commands coming down from the top, and that is doubly true when those orders concern food and drink, something people feel deeply personal about. Beginning in infancy, what human doesn't react with pursed lips, a shaking head, and a tight scowl when told what they must or must not ingest? People's routines around food play a big role in who they understand themselves to be. Abolishing a decades-old custom and snatching away that perk must have felt like a personal attack.

This is as true for the country as a whole as it is for White House workers, and it demonstrates the challenges and complications our nutrition efforts faced. Americans would never change their eating habits if doing so was perceived as submitting to a directive from the nanny state. Changing entrenched foodways will only work if the culture is ready, and my colleagues clearly were not ready to forgo those Cokes. If you try to make changes and the culture is not supportive, not only will you not get what you want, but you will be weakened the next time you attempt to introduce something. It's not unlike picking fruit. You have to wait until it's ripe. This applies as much to local scout packs and church suppers as it does to the White House.

## SEEDS OF CHANGE

Even as the soda debacle was unfolding, we were building a program that exemplifies the correct way to change the culture's attitudes toward food. I'm talking about the White House Garden.

The idea to plant that garden arose during one of those conversations Michelle and I had in the quiet of her Chicago kitchen after dinner had been cleared and the girls had disappeared into the nether regions of the house. Once we came to Washington, planting a garden seemed like a perfect way for Michelle to ease into addressing food issues without political risk. After all, who doesn't love a garden? What type of person is going to condemn a well-meaning mother for wanting to cultivate a small rectangle of ground to nourish her kids with fresh fruits and veggies?

Legislators, executives, and local leaders love nothing more than trial balloons: little plans that allow them to float an idea to gauge impact and avoid potential backlash from proceeding too quickly in an unpopular direction. By establishing a garden for all to see, we would get a sense of what the culture felt about the First Lady taking on food policy as an initiative. And if all went well, then Michelle could start engaging in more controversial issues, knowing that she had planted some public goodwill.

"We're going to do it, and if it goes well, we will start the work of launching a national initiative on kids' and family health," Michelle said, sealing the deal.

Yes, the garden would provide tomatoes, spinach, strawberries, and (of course) kale, but Michelle and I also realized that it could yield a cornucopia of other, less tangible harvests. It could be a classroom for inner-city schoolkids while sending a gentle message about the sorry state of what the nation's children were forced to eat in cafeterias, no sanctimonious preaching required.

The First Lady wouldn't have to utter a negative word. The garden's very existence would be in effect an advertisement for fresh, healthful, organic produce broadcast from one of the highest-profile places in the land. It would be a tangible expression of our values when it comes to food and our children. Its quiet message would be a shot across the bow to the callous food and beverage company executives who profited from foisting unhealthy junk on the nation's children, much of it derived from mono-cropped cornfields and other environmentally destructive agricultural practices. All that she'd have to do was kneel down and pull up a bright orange carrot, smiling radiantly at the cameras. It would be a highly visible way to let the world know that as far as food policy went, a new sheriff had come to town.

"I planned to use the work we did in the garden to spark public conversation about nutrition, especially at schools and among parents, which ideally would lead to discussions about how food was produced, labeled, and marketed and the ways it was affecting public health. And in speaking on these topics from the White House, I'd be offering an implicit challenge to the behemoth corporations in the food and beverage industry and the way they'd been doing business for decades," Michelle wrote, in her memoir.

## SACRED GROUND

But as with everything in Washington, wishing for a White House Garden was a lot easier than actually sinking a spade into a patch of dirt. Michelle delegated that task to me during the administration's early months. First of all, understand that we proposed to dig up part of the White House lawn, one of the most groomed and pampered patches of turf in the world and, in addition, one of the most well known and often photographed. Home at the time to Jackie Kennedy's legendary rose beds. Site of impromptu presidential press scrums. The grounds have always been considered the property of the people, not a place the current, temporary occupants of the big house located thereon could deface on a whim. Getting upset at changing this American icon is one of the few truly bipartisan things left: Look at the yowls of complaint that greeted Melania's redesign of Jackie's flower garden.

Fittingly for a national treasure, the grounds were maintained by the National Park Service under the uncompromising watch of Dale Haney, who had been on the job since 1972. (He doubled as the unofficial friend and chief walker of a small pack of beloved First Pooches, from Richard Nixon's Irish setter, King Timahoe, to Obama's Portuguese water dog Bo and Biden's German shepherd Commander.) Although Haney and I eventually became great friends—I truly love the guy—his first allegiance went to the precious eighteen acres entrusted to his care. His round, jovial face fell when I told him that we intended to plunk a bunch of raised beds on his territory. He was first and foremost a dedicated National Park Service employee, and his grimace wouldn't have been deeper had I just proposed putting a hot dog stand directly in front of Old Faithful.

He responded with a tentative "Huh? I don't know if I can do that."

"Maybe you and I have to figure it out," I said. "The First Lady really cares about this project."

He exhaled loudly. "Okay, let me work on some options for where you might be able to plant it."

A few days later he approached me wearing a relieved look and reported that he had found a perfect spot. He led me down across the South Lawn and past the Oval to a space about the size of a Ping-Pong table hidden in the deep shade between a tennis court protected from public view by vines, trees, and a metal maintenance garage that housed tractors, lawn mowers, trimmers, and other mechanical devices necessary for keeping the grounds immaculate. To Haney, the site had the virtue of being tiny and out of sight, but horticulturally it was easily the worst place on the property to try to grow a garden.

I said, "I don't think you understand. Come with me," and strode out into the middle of the South Lawn. "How about here?"

He stuttered. "Uh, no. No way. It'd be smack in the middle of one of the most iconic pictures in the world all the tourists like to take of the White House from the street."

"Exactly!" I said. "Michelle wants it to be highly visible."

"I have to talk to some people."

The next day he proposed a compromise. It was a sunny enough spot farther down the lawn and a little to the left side. It would not appear in the middle of that popular photograph but be clearly visible to the throngs of visitors who gathered daily on the E Street sidewalk. He agreed to dig up an eleven-hundred-square-foot area, a fine start. But just a start. Over ensuing years, we managed a few more land grabs and gradually doubled its size.

Michelle was no gardener. Born and raised in Chicago, she never had been. But she was determined that the food she would feed her family would be a message about caring for the environment as well. Her crops would be grown organically without the help of artificial fertilizers and pesticides, even though reaping a successful harvest would have been a challenge for someone with a far greener thumb and a full agrochemical arsenal.

Nonetheless, on March 20, 2009, she got down on her hands and knees in freshly cultivated earth beside a couple dozen Washington, D.C., fifth graders, who probably had more growing experience than she, given that they were veterans of their own schoolyard's garden. I was on hand, as was Secretary of Agriculture Tom Vilsack. As the former governor of Iowa, I assumed he knew something about getting his hands dirty. It was an electric and magical day, one that felt surreal at the time (and still does to this day). All of the conversations back in the kitchen in Chicago had begun to manifest themselves. I tried to stay present in the moment and take in the significance of it all as we dug up one of the most historic lawns in the world.

At the same time, we all felt some trepidation as Michelle and her young assistants pushed those tiny seedlings into the ground. "This better work," Michelle whispered to me, widening her eyes meaningfully. I thought to myself, "Yeah, no shit! I am aware that this better work!" Fortunately, our garden flourished from the outset, with fulsome support and plenty of expertise from Haney and the White House horticulturist Jim Adams. Later that spring, the same group of kids returned, this time to harvest early crops and bring them into the kitchen, where we made lunch from the first produce to grow in White House soil in more than fifty years.

From its inception, I felt in my heart that the garden was going to be a big generator of positive publicity, far bigger than most members of the administration realized. The teams in both the East and the West Wings largely underestimated what was taking place on the South Lawn.

But even I was blown away by the intensity of the response. Positive media coverage not only blanketed this country but resonated around the world and got louder every time cameras came to capture yet another photo op in the sunshine starring Michelle, schoolkids, and her lush, healthy plants. The act of growing food to nourish the next generation is a deep expression of the human condition. As I saw the photos on front pages of newspapers around the globe, from China to Afghanistan and dozens of other countries, I felt the power of food to connect us and remind us what we have in common.

Gardens have a way of taking on a life of their own. The one thriving on the South Lawn was no exception. By summer, we were producing hundreds of pounds of produce, more than fifty different varieties, enough to supply ourselves and donate excess to Miriam's Kitchen, a D.C. food charity. At the risk of sounding like one of those greener-than-thou food snobs, I hereby admit that on most afternoons I removed my senior adviser's tie and suit coat, slipped into a chef's jacket, and meandered down to the garden to see which vegetables and fruits were at peak ripeness. They became the central ingredients around which I built the nightly dinners I prepared for the Obamas.

But my daily strolls immediately became more than mere harvesting trips. The garden became my little oasis, a sanctuary from the verbal fisticuffs with politicians, lobbyists, and often fellow staffers that took up too many of my working hours. It reminded me that there is value in taking time out to appreciate what you're fighting for, to allow for a little self-care. Sometimes I'd wander down just to get a break from the swirl of the White House. Michelle felt the same way. She'd go down on the weekends when the place was quiet, watching the plants grow, pulling the occasional weed, or simply unwinding. One time I rushed down to gather some ingredients for dinner and came upon Barack and the German chancellor, Angela Merkel, strolling together among the raised beds after a daylong summit meeting, checking out the plants and sneaking a couple nibbles.

There were occasional reminders that this was the White House Garden, not your average backyard plot. We had strict instructions to inform the Secret Service before we went there. When working the soil, we had to be careful not to disturb any of the subterranean sensors that would alert security to the footfalls of possible trespassers. Our growing veggies could not block the views of surveillance cameras. On one occasion—which I found out after the fact, thank God—a rooftop Secret Service sniper had my head in the crosshairs of his telescopic sight while I was gathering ingredients for dinner.

Inconveniently, I'd gone there unannounced when someone had jumped the fence and was on the White House grounds without authorization. Not knowing there was an issue, I gathered my produce and returned to the White House to cook dinner. As I was heading upstairs to serve it, one of the guys on the Counter Assault Team, a giant dressed in all black with his large, long-range rifle on his back, emerged from the elevator. "I had you scoped, thought you were the jumper. Was a close one," he said to me. Needless to say, I told the Secret Service every time I went down to the garden after that!

## A WORLD-CLASS PR FLUB

Alas, I learned that even the most benign actions in Washington have their detractors. Upon learning about Michelle's horticultural ambitions, the Mid America CropLife Association, which despite its down-home name is a trade group for such chemical pesticide giants as Bayer Crop Science, Dow AgroSciences, and Monsanto, fired an open, three-page diatribe at Michelle not so gently upbraiding her for growing organically:

> As you go about planning and planting the White House garden, we respectfully encourage you to recognize the role conventional agriculture plays in the U.S. in feeding the ever-increasing population, contributing to the U.S. economy, and providing a safe and economical food supply. America's farmers understand crop protection technologies are supported by sound scientific research and innovation.

In case there was any doubt about the true motives behind CropLife's letter, someone soon leaked an internal email sent by the organization's director, Bonnie McCarvel:

*Did you hear the news? The White House is planning to have an "organic" garden on the grounds to provide fresh fruits and vegetables for the Obama's [sic] and their guests. While a garden is a great idea, the thought of it being organic made Janet Braun, CropLife Ambassador Coordinator, and I shudder.*

I could not have been more delighted with the industry group's ham-fisted letter, which still ranks among the great, world-class PR flubs. It only bolstered the garden's green, organic, and environmentally friendly message. More important, chemical agribusiness's misstep showed that we were doing something worthwhile. Our garden was sending a powerful message to friends and foes alike. We had put a big stake in the ground—literally—and could gauge the electorate's reaction. The overwhelmingly positive results allowed us to bank vital political capital.

We needed all the capital we could get. When we took office, most of the food issues we cared about were far out on the sidelines of mainstream culture. True, a small, vocal, fragmented group of good-food advocates and industry had been talking—mostly at each other—for more than a decade, but the average voter was not focused on food policy at all, and certainly not voting on that basis. It was far from being a motivating political issue. Quite simply, the vast majority of Americans remained dispassionate about food issues. They just wanted to satisfy their hunger with something good, fast, convenient, available, and affordable.

With the garden, we tried to take the conversation about better food to where the people were emotionally and to engage them in a way that they understood with something upbeat. The White House Garden helped elevate a different set of cultural values and gave people an opportunity to express their support for it. That gave us the license to push our agenda forward and to say that as a nation we all needed to do a better job with nutrition. Beyond that, we could achieve this goal by raising foods in an environmentally friendly manner. The garden seems normal now, as if it belonged and always should have been there. But tearing up the most iconic lawn in the country and planting a bunch of eggplants, chard, and peppers were radical things to do and helped show the way forward. The fact that it does feel normal now is a sign that our culture has changed.

Don't let the wholesome symbolism lull you into the impression that the garden was just a mere PR gesture. Far from it. Like many of Michelle's outwardly fluffy and goofy stunts, its mission was strategic and laser focused. Along with those eggplants and tomatoes, the garden gave us license to press forward with our goal to improve the food system. It helped us expand Let's Move! Over the next seven years, the administration was able to increase access to SNAP benefits. Barack, dealing with an economy in free fall, nonetheless went to work with Michelle to persuade Congress to approve and overhaul the school lunch program, which included free breakfast and lunch to every schoolchild in less affluent districts. We made changes to food labeling laws, added fresh fruits and vegetables to the Women, Infants, and Children program, banned trans fats, worked to limit food marketing to kids, curbed antibiotic use in livestock, and put calorie counts on chain restaurant menus. All of these victories took root that first spring in those raised beds on the South Lawn.

## LATE NIGHT WITH MICHELLE

In early 2012, the following teaser appeared on the First Lady's Twitter account: "[Jay Leno] says he never eats vegetables . . . we'll see about that on the Tonight Show."

It was true that Leno, who hosted *The Tonight Show* on NBC for twenty-one years, maintained a terrible diet. He once admitted to an interviewer that he lived off junk food: pizza, hot dogs, and hamburgers washed down by plenty of soda—not the sugar-free kind. He claimed never to have eaten a salad and that the last time a vegetable had passed his lips was in 1969. The most recent fruit, an apple, in 1984.

"I have nothing against vegetables," Leno told Michelle on the set of *The Tonight Show* after she took the seat beside his desk. "They seem like very nice plants."

She presented him with a whole-wheat pizza topped with tomatoes, sweet peppers, eggplant, and zucchini. Leno ate a slice, allowed that he liked it, and said that he assumed it was pepperoni and sausage. The segment's punch line came when Leno presented Michelle with a plate of sliced beets, about the only vegetable she dislikes. Nonetheless, with a look of horror, she gamely nibbled a few bites.

Michelle, who constantly reminded us that she wanted her name recognition and popularity to be put to the best possible use, became known for connecting similar lighthearted media moments to the serious message of better nutrition. At the White House, she outmaneuvered the legendary basketball star LeBron James to make a successful dunk. After his "humiliation" they sat together at a table, each munching an apple. James, who stands six feet, nine inches tall, weighs around 250 pounds, and is a role model for millions of young people, explained to her that eating healthy helps him stay on top of his game. Jimmy Fallon, dressed as a ditzy teenager, confronted Michelle's offer of kale chips with a long, exaggerated "Eeeeew." Michelle did push-ups with Ellen DeGeneres. She joined NFL players for a few downs of flag football, demonstrating considerable hand-eye coordination as a receiver. The traditional White House Easter Egg Roll became a mini health food fair. Kids in attendance had plenty of fruits to hunt. They planted vegetable seeds in paper cups to take home. Celebrity chefs gave cooking demonstrations. Working with his grade-school-age granddaughter, Jacques Pépin showed how to make crepes stuffed with chard, spinach, and thyme just harvested from the White House Garden. The First Lady's Kids' State Dinners became annual events. Children from all the states and territories as well as Washington, D.C., submitted entries in a healthy recipe contest. The winners from each area came to the White House for a celebratory meal, joined by the president himself. The eight-to-twelve-year-olds got the full state dinner experience, just like heads of state.

Michelle's list of hijinks went on. It might at times have looked like fluff, but we carefully planned every appearance so that it kept nutritious food in the national spotlight. We wanted our campaign for better food to be warm, inviting, and lighthearted. Better to have a beautiful, smiling woman than a scowling Uncle Sam at the head of their dinner tables.

We also realized that if we wanted people to change, we would have to meet them where they were. We would have to find ways to get kids and families engaged and help bring them to a different place with food. We would have to work to infuse the culture with a different set of values when it came to our health and how we feed ourselves. We would come to them, not insist that they come to us. We would never say "You should eat the way we do," a message blared out ad nauseam (and to little effect) by socially conscious foodies. But, hey, if you make eating well seem like something fun on Leno, or if you have LeBron eat an apple beside Michelle, you can painlessly inject the values of good nutrition into the culture.

## MY LAST SODA

I can imagine that you're thinking something along the lines of this: An attractive, beloved First Lady may be able to make use of her natural gifts and high visibility to promote change on a macro level, but what about the rest of us? My answer is that most cultural change does not take place on the national level. Local leaders, chefs, teachers, religious leaders, moms and dads, and all of us can and do make small changes that can become the basis for much larger trends. You can start at your own dinner table.

I'll cite my childhood friend Keronn Walker as an example of how one person can change culture. Although decades have passed, I vividly remember the last time that I drank a sugary soda. When I was eighteen years old or so, Keronn and I had stopped for gas at a convenience store, and I dashed inside. He was more than a friend. Keronn was someone I looked up to intensely. We both played college baseball with a shared dream of making it to the major leagues. Not only was he a couple of years older than I, but he was a hell of a lot better ballplayer, soon to be drafted by the Kansas City Royals. When I got back in the car holding a bottle of Mountain Dew, he shook his head. "What are you drinking that shit for?" he said. "It's poison. Look at the label. See all the sugar and chemicals? Don't drink that stuff."

That's not how I would write an ad for the general public, but for me, an aspiring teenage athlete hanging with a friend I admired, it was the way that message got through. It was the language, the way it was said, sure, but more important, it was the relationship between us. I trusted Keronn, we had been close for years, he said his piece, and that was that. I looked down at the bottle, read the label, said, "You're right," and haven't had a soda since.

I'm telling you this because it's an example of how every single one of us can change food culture. Most of us will never come close to having the bully pulpit and media magnetism that Michelle Obama and the celebrities who joined her deployed so effectively, but we can all make changes in our own lives, in the lives of our family members, and in the lives of friends. We all impact the norms, attitudes, and behaviors of those around us. This is true in our homes, workplaces, houses of worship, and friendships. We have begun, and must continue, the long, slow work of resetting the norms of what it means to nourish ourselves. Small changes can create positive feedback loops. Eventually a cultural force builds that—who knows—might someday influence the food attitudes of administrators and politicians at the highest levels in the country. Just ask Keronn.

## WATERMELON REVOLUTION

If you really want to see the cultural power of one person's grassroots leadership and how that can balloon through the entire food system, a good place to look is Burke County, a part of eastern Georgia where two-thirds of schoolchildren come from homes receiving SNAP. In 2010, the wife of an area farmer approached Donna Martin, head of Burke's school nutrition program. The farmer had a problem, and his wife wondered if Martin could help. Her husband's fields had produced a bumper crop of watermelons, far more than the usual customers wanted to buy. Could Martin help? Martin said that the kids would absolutely love the melons and offered to buy them all.

Martin, normally a friendly extrovert, can be bullheaded when it comes to feeding "her" kids—all five thousand of them. Her motto is "I'm going to do whatever it takes to get the job done." Word that she was happy to buy farmers' excess produce started to get around the local agricultural community, and other growers approached her offering okra, tomatoes, collards, butter beans, black-eyed peas, salad greens, new potatoes, and locally grown and ground grits (as opposed to the instant kind, which she describes as "horrid"). The students accepted the changes, and Martin saw an opportunity to transform the food served in her cafeterias. Pizza, French fries, fried chicken, canned vegetables, and white bread would be banished and replaced by fresh, local, and whole ingredients.

At first, she almost got run out of town. Parents and teachers complained. "The students won't eat that." "They want fried chicken." "They hate brown bread." She was told that her cockamamie notion would result in hungry kids and dumpsters full of kitchen waste.

"It's tough to buck the system," she told me. "People around here like greasy, sweet food. But I kept my head up high and said, 'I'm a registered dietitian. This is what is good for the children.'"

And a funny thing happened. The students loved the new menu. They enjoyed two main courses each day and could choose from five fruits and vegetables. Teachers, who had been skeptical if not openly hostile, became converts. Their zeal created an unforeseen problem for Martin because the educators began to ambush the farmers' delivery trucks as soon as they pulled in to the parking lot outside the kitchen doors, clamoring to buy produce meant for the kids to take home for themselves. Making a virtue out of this problem, Martin catered to the demand she had created by starting a farmers market in Waynesboro, the Burke County seat.

About five years into the new program, a principal asked to speak to Martin. He had something important to say. She shuddered, sure that he was about to complain. "Here it comes," she thought. "He wants fried chicken and hot dogs back on the menu."

Instead, he said, "Donna, initially, when you made the changes, I wasn't in favor. I wouldn't eat the meals, but now I see a difference in the kids' academic performance and behavior. It's changed how I eat, too."

School nurses reported that student complaints of stomachaches, once common, had all but disappeared. At one time, many kids had no breakfast. Now they all had access to a nutritious meal at school before classes began. Test scores went up. Attendance rose. Tardiness plummeted, particularly after Martin instituted a policy where the first three hundred students to arrive each morning got free fresh fruit smoothies.

Wanting to capitalize on the trend, Martin had the children grow some of their own food in schoolyard gardens. Excess went to charities. She introduced cooking classes that emphasized healthful ingredients, making sure that each student went home with a printed recipe for that day's dish, which could be as simple as oven-roasted cauliflower (if you haven't cooked cauliflower that way, you're missing a treat, even if you don't think you like the vegetable).

In an area where extended families are the rule, parents, grand-parents, uncles, aunts, and siblings all began eating healthful fare at home. After Martin served kiwi at one meal, parents called her demanding to know the name of that "fuzzy brown fruit" their children liked so much. Area grocery stores began adding not only kiwi but mangoes, jicama, and sixty other produce items they previously had not stocked.

While shopping one day, Martin overheard a mother asking her seven- or eight-year-old daughter, "What the heck are those things?" pointing to the produce display.

"Star fruit," said the little girl, who ate them in the cafeteria and liked them.

Says Martin, "If they grow it, if they cook it, if they taste it, they will eat it."

## FRIDAY-NIGHT MIRACLE

What ultimately convinced the entire community that Martin was onto something truly big occurred on the high school football field, not the cafeteria or classroom. "Football is king around here," Martin said. "That's the only thing going on Friday nights in the fall."

Sadly, the Burke County Bears had a history of disappointing their loyal fans, usually languishing in the middle ranks for their region. But in 2011, they not only won their regional tournament but took the state championship against the once-invincible Peach County Trojans. When asked to explain the team's success, Coach Eric Parker didn't attribute it to practicing harder, more strenuous weight-training sessions, or executing new, clever plays. He credited it to a single factor: The team was eating more healthful food at school.

The successful and popular coach, one of the most influential members of the tight-knit, largely rural community, let the audiences of local radio stations, newspapers, and TV channels know that congratulations for the Bears' victory should go not to him but to the lunch ladies. Those hard-assed football players, respected by all for their strength, speed, and agility, owed their greatest victory to proper diets. In Burke County, healthful food became cool.

Jaylon Kelly, who went on to attend Georgia Southern University after high school, played on the offensive line for the Bears. He admits that he was not won over by healthy food until his junior year. Like many youths, he disliked vegetables. His lunch sometimes consisted of a bag of potato chips. Breakfast was something he, like most of his friends, never ate. "I had to get up at 5:30 to catch the bus and didn't have time," he said. But he found that eating three regular, healthy meals a day at school (Martin had by then also started an after-school dinner program) dramatically improved his on-field performance. "It really showed. I could do so much more."

He put himself through college by working summers for Martin and continued eating a nutritious diet, even though the offerings at the college dining halls didn't always live up to Martin's standards. To avoid fatty red meats and anything deep fried, he sometimes had to resort to vegetable-forward Chinese food two or three times a week. "I ate healthy every day," he said. "Or I tried to. I'm only human."

Stories like the one out of Burke County provide more ammunition for government officials wanting to improve school food than all the polemics and op-ed diatribes ever published. What's not to love about Martin's program? No politician is going to come out and say they are against doing what's best for children's health. Even tight-fisted Georgia good old boys approved of actions that boosted academic performance by 17.5 percent and increased graduation rates by 20 percent—and, better yet, made all those Friday-night, vote-casting football fans smile.

The First Lady picked up on the story of Burke County, personally visiting Martin to help the kids harvest the crops they had grown. As always, Michelle was accompanied by a media scrum. The resulting coverage made better school food a national sensation, and the trend spread to other school cafeterias across the country. With the cultural winds blowing strongly in our favor, we were able to push hard for the federal Community Eligibility Program. It enabled all schools where 40 percent or more of the student body came from low-income homes, as did most of those in Burke County, to serve breakfast and lunch to every kid free.

Prior to that, only underprivileged children got free breakfasts, which created such a painful stigma that many skipped the meal rather than being labeled poor. Giving every student free meals regardless of their family's financial status removed that stigma. Riding the same cultural wave, the Obama administration's Healthy, Hunger-Free Kids Act mandated more nutrition options across several government initiatives. We empowered thousands of Donna Martins in school systems across the country.

By influencing culture, we made inroads into sectors that I

thought were all but impervious to our better-food message. In 2009, the first time I attended the annual convention of the School Nutrition Association, whose fifty-five thousand members provide meals for more than fifty million American schoolkids, I honestly started weeping. As I entered the vast exhibition hall, tears literally rolled down my checks as I looked out over booths belonging to businesses that sold every imaginable type of junk food. These companies shamelessly promoted garbage to the people who fed our kids—think of that. As a parent, you put so much trust in the school where your kid spends much of their waking hours when they're not at home with you. And here, the folks who feed those kids were inundated with messaging promoting every hyper-processed food imaginable. When I returned to the same gathering a couple of years later, the hall fell far short of what any of us fighting for better food would hope to see, but I noticed some bright spots—sweet potatoes, whole-grain products, and other unprocessed offerings. The progress was definitely visible.

Unexpectedly, I encountered a group of supermarket executives I had sometimes had political confrontations with. I asked why on earth they were prowling the aisles at a school nutrition trade show. After all, the attendees bought their cafeteria supplies from food-service companies, not supermarkets. The corporate guys said that they had begun to have serious problems keeping certain produce items in stock. "Cauliflowers!" one exclaimed. "Suddenly we have a run on them. Another week, it would be butternut squash. Or something else. We couldn't figure out what was going on."

The cause for the unexpected demand spikes was that kids were coming home from school and telling their parents about a new vegetable they'd had for lunch and loved. For leaders of huge national grocery chains, hanging out with several thousand school chefs had become a way to get a preview of what the next in-demand vegetable might be.

## THE POWER OF A CHEF

Marketers will tell you this: When you are trying to shift culture, seek out the influencers—those who occupy positions that enable them to be outsized opinion leaders, be they athletic superstars or local football coaches. Then target your messaging through them.

I'd already seen glimpses of the influential power that chefs can have over the broader food culture through what they put and do not put on their menus. Early on, chefs had the nation passing on Chilean sea bass and giving swordfish a break, long before conserving overfished species became national movements. Unless you were familiar with the cuisine of the Andes, you'd probably never heard of quinoa before chefs, attracted by its nutty flavor and novelty, started serving it. Now it's hard to pass a salad bar without encountering the South American grain. Kale's place at the table was pretty much relegated to that of a frilly, visual garnish; the only place you'd likely find it in the supermarket was as a decoration somewhere. Then chefs started using it. Between 2007 and 2012 its appearance on restaurant menus increased by 400 percent. The amount grown in this country shot up fourfold. Today, it's nearly impossible to find a grocery store that does not carry it. The farm-to-table movement did not exist here until the early 1970s, when Alice Waters and a few other Bay Area chefs popularized it. The word "locavore" first came into our lexicon as recently as 2005, when Jessica Prentice, a chef, of course, coined it. The cultural value of sourcing local now permeates the chef world, with many of the biggest food-service companies and retailers following suit. All of these changes swept through the culture with incredible speed.

Now, imagine what could happen if chefs shined their spotlight on other underutilized foods that are nutritious, accessible, and affordable and have a low environmental impact. Researchers working with the World Wildlife Fund and Unilever have simplified finding such foods by compiling the Future 50 Foods list, a roster of whole grains, legumes, and vegetables that are good for both the planet and diners' health. Chefs have the power to make them tasty and popular. And if you like, you can consult the list for the meals you serve your family.

## UNTAPPED POTENTIAL

Despite such accomplishments, I'm convinced that chefs have deployed at most 5 percent of the power they have to change the cultural landscape around diet. The phenomenon of the celebrity chef has provided a tremendous megaphone. Every person in power goes to restaurants—every senator, every congressman, every school board member, every small-town mayor. Often, they know their favorite chefs personally and have relationships with them. Lobbyists and other influencers spend piles of money (much of it in restaurants) to gain a fraction of the access chefs have. A media megaphone and the right connections are a recipe for outsized influence.

Chefs have long been famously charitable. They are always donating their time and resources to support worthy causes. As soon as the last tremors stopped, José Andrés was in Haiti setting up mobile kitchens and has since become a global icon for his work with World Central Kitchen. The New York chef and *Top Chef* judge Tom Colicchio has been a force for hunger relief programs, as is Atlanta's Asha Gomez. Dan Barber leads the Stone Barns Center for Food and Agriculture, an educational organization. Alice Waters is behind the nationwide Edible Schoolyard Project. Renee Erickson, owner of seafood-forward restaurants in the Seattle area, has successfully lobbied for marine conservation programs. Chef Sean Sherman, who grew up on the Pine Ridge Reservation in South Dakota, works to highlight once all-but-forgotten traditional indigenous foodways. Michel Nischan started Wholesome Wave to bring access to fruits and vegetables to low-income communities. Bryant Terry of Oakland works on issues of food justice for children in underserved neighborhoods. Chefs everywhere donate their talent to improve life in their communities, but so far we haven't wholeheartedly addressed the issues that will in the future have enormous impacts on what they will have available to cook. They should work to save the planet just as they saved swordfish.

When we launched Let's Move!, the first group of allies we turned to were chefs. It was an inspiring sight to see more than five hundred of my peers gathered on the South Lawn dressed in white jackets as Michelle introduced them to a program we named Chefs Move to Schools, an effort to have cooks from across the country get involved with their local school nutrition programs.

We left it up to the chefs to design activities that fit their circumstances. Some acted as unpaid consultants to cafeteria workers, giving them suggestions on how to make their meals more exciting for kids while still staying within their notoriously meager budgets. Others worked directly with kids. In Orlando, chefs and students collaborated to develop new menu items, giving the kids a sense of pride and ownership, and of course the students devoured their creations. Paul Kahan, a highly respected and beloved Chicago chef (and my former employer), launched a program called Pilot Light with the goal of enhancing young people's perception of food through hands-on teaching.

It was a good start, but it failed to amount to a transformation of the role of chefs. If we are going to harness the 95 percent of chefs' power that lies dormant, we have to come up with a series of clear priorities. We need a cohesive national strategy for chefs to organize around. The most obvious place for most to begin is on the plates that come out of their kitchens. Although harried cooks might not have the time to reflect on it, every restaurant is a teaching institution where guests learn culinary norms. Too often the lessons are negative: A big piece of meat in the center of a plate equals value. But a chef can also create another norm, perhaps by serving a delicious dish of beans and escarole that creates a new understanding: that a meal of these kinds of foods is exciting and satisfying. These are lessons that customers take home with them. Better-for-the-environment menu selections also provide economic opportunities for climate-friendly producers by giving them an economic outlet, shifting money from "bad" food to "good." I call it procurement power.

What ends up on people's plates shapes our food norms and culture. If diners get dishes that are mostly meat, with vegetables as a small accompaniment, they learn from that. If vegetable-forward dishes become what dinner looks like in the minds of restaurant clientele, they take that back to their homes. I am continually shocked, even in the face of a tsunami of climate catastrophes playing out around us, that chefs—many of whom are enlightened friends of mine—continue to open steak houses, for example. Again, I don't want to be holier than thou about this. I love a good steak, even if I reserve it for very special occasions. But I find it unconscionable to keep promoting a massive piece of beef as what "dinner" looks like. On the flip side, chefs can teach us with every dish. They can show us what a modern plate of food should be and establish the norm that meat should take a supporting role most of the time. They can teach us how to make an Earth-friendly future delicious.

Chefs also educate customers by telling stories. Menus, waitstaff, and social media can all tell stories. "Local" is a great example of how this works. Two decades ago, you'd never see the word on a menu. Today—sometimes to a fault—it's everywhere. Waitstaff tells the stories of farmers and other producers behind restaurant fare and explains the benefits of eating locally. The same holds true for sustainable seafood options. It isn't a stretch to see that sustainable producers and products receive the same treatment. A chef can truly help customers understand food and climate, as I tried to do through my Last Suppers. The message should be this: If you enjoy this food and want to continue eating it, you should also care about climate change.

The role that chefs play in setting trends and shaping norms, impacting the supply chain, and telling the stories that make it all make sense gives chefs a unique understanding and credibility with key influencers and policy makers on food issues. Chefs can significantly influence food policy by leveraging their public platforms, expertise, and leadership in the food industry. Although there are amazing leaders doing great work, we have only scratched the surface of our potential. From my vantage point, our top priority as a community should be on climate change and environmental health. Whatever other critical issues a chef is working on—food justice, hunger, farmworker rights, to name a few—all of them will be upended and exacerbated by climate. I am not saying abandon those critical problems that need urgent attention, but we all need to be pushing for more aggressive climate policy while shaping our restaurant operations to help catalyze a transition to a more regenerative food system. In my view, chefs need to connect their causes and all the suppliers that make their work possible to the larger issue of climate, and how food and agriculture can help stave off the worst effects of climate change. Our entire industry—not to mention our way of life—depends on it.

## HOW TO HIJACK A CADILLAC

While it's true that shifting cultural attitudes is hard, practitioners of one profession have shown themselves able to do so time and time again. They have developed a sleek tool kit that can be used to alter the direction not only of food culture but of fashion, entertainment, how we vote, and even cigarette smoking and consumption of alcoholic beverages. In addition, this same profession has developed techniques to use cultural changes to drive concrete results. I'm talking about the sophisticated marketers who are busily changing your personal habits even as you read this, and you probably don't even know it.

Beginning after World War II, by popularizing mass-produced foods developed for the military, consumer marketers helped foster a generational shift in how Americans dined. Women whose mothers, grandmothers, and great-grandmothers had generally viewed feeding their families as a duty steeped in traditional culinary practices were suddenly joining the workforce in unprecedented numbers. Coinciding with the dramatic shift in the role women were playing in the workforce, a nascent but growing food industry widely promoted the message that cooking was pure drudgery, best avoided by serving packaged, canned, and frozen fare that could be popped into the oven and put on the table in minutes. It is true that these convenience foods helped bring women newfound freedoms. No longer shackled to their stoves, women were able to pursue careers they found more meaningful and rewarding. This liberation was incredible progress, to be sure. But it ultimately also led to an unintended consequence: Eons of culinary tradition and skills disappeared within a few decades, and a generation of Americans grew up not knowing how to cook. Having achieved this cultural shift, marketers stepped in and shifted our culture to value unhealthful processed meals, fast food, and TV dinners. And over the decades, they got better and more sophisticated at shifting our behaviors to consume larger and larger amounts of their ultra-processed foods.

But what if they had instead used their transformational tool kit for the good? At the White House, we decided to put that question to the test. If it worked, the ramifications could be huge.

## CHANGING SIDES

After nearly a decade as a senior executive at the Nielsen Company, a worldwide market research and media measurement organization that most Americans know through its Nielsen television ratings, Karen Watson was becoming disenchanted. Food and beverage advertisers, many of them Nielsen clients, used their enormous promotional budgets and huge stockpiles of consumer data to aim their pitches for junk food and sugary drinks at less affluent communities, particularly those that were predominantly Black or Hispanic. A diet filled with highly promoted, unhealthy products set these segments of society up for a lifetime of obesity and chronic disease.

Watson had observed that the government agencies trying to disseminate messages to improve public health were badly outgunned on many fronts by the firms pushing junk food. Nutrition advocates lacked the necessary data to accurately target their campaigns. They had no coherent and consistent messaging. Their approach often did not adhere to the principles and strategies that change attitudes and behaviors. They failed to measure the results of their efforts to determine whether they actually delivered—and to make necessary changes if they fell short. What would a public health campaign look like if government and nonprofits had industry's tools and deployed them to push healthy and environmentally friendly products instead of ones that make consumers sick? Aware of the administration's interest in better food publicity generated by Let's Move!, Watson arranged to meet with us.

We got together in early 2013 when the administration, fresh from winning a second term, felt strong and in a position to make bold moves. Watson volunteered to give us a thorough schooling on modern marketing—from research to final results. We jumped at the opportunity, and in short order she cajoled more than thirty of her high-ranking colleagues from different divisions of Nielsen to share their consumer data with us at no cost. Most did so willingly, if not out of enjoying a civic-minded break from foisting potato chips and soda pop on Americans, then because President Obama had publicly thanked Nielsen's chairman for his company's help. (It was the first the boss heard of this employee's unofficial volunteer work.) The chairman in turn expressed his appreciation of his workers' civic-mindedness.

When you give it some thought, soft drink and processed food companies are actually as much in the marketing business as they are in the food and beverage business. Many of the biggest food companies don't actually manufacture a portion or the majority of their products; they outsource that to co-manufacturers. Marketing, more than quality or value, differentiates their products from competitors' and builds customer loyalty. Any attempt to control processed food companies' promotional efforts aims directly at the heart and soul of their business model. Understandably, they defend themselves ferociously.

Until Watson came along, we'd merely tinkered around the edges of Big Food's efforts to persuade the public to consume their less-than-good-for-us brands. Marketing is considered free speech and is entirely protected under the First Amendment; you can't ban companies from selling their products. But maybe we could use marketing to our advantage. Given its importance in driving consumption, I was looking for strategies to move the needle. I told Watson that I was eager to launch a program that would give us a clearly measurable win. One of Michelle's standing goals had been to reduce soda consumption, particularly by kids. We had been hammering home the message to eliminate sugary drinks, including making it a core message of the Dietary Guidelines released by the government every five years. But what if we fought fire with fire? Together with Watson, we agreed to focus on trying to persuade the public to consume more water.

There might seem to be a disconnect there. Why promote water consumption when your goal is to reduce soda intake? Karen had a core piece of data: The more water people drank, the less soda they would consume. The second reason was obvious once we understood the nature of marketing. Following the model of virtually all effective communication programs, our message would be totally positive and look to make an emotional connection. Pepsi's ads never tell people not to drink orange juice. Instead, their promotions make sipping Pepsi appealing by wrapping the action in images of happiness, friendship, and love. The science and insights that these companies had from Nielsen and others was that the best way for your advertising to trigger a behavioral change was to elicit an emotional response, and even better if you could simultaneously trigger a memory response. Showing your product in a light that made people feel love or think about their kitchen table when they were a kid sold.

Yet public health advocates had always focused on education, not emotion. Teach people why healthy options are good for them and what a good diet does to the brain! But that's an ineffective way to get people to change. Love, sex, and happiness beat fiber, vitamin B, and beta-carotene every single day.

So, we were going to give water all of the marketing oomph that soft drinks receive. Water would become a hero. Drink water and you'll feel great, be in terrific physical shape, look good. We had no need to go negative on sugary drinks, and in fact actively avoided doing so because sophisticated tests showed that was not an effective approach.

One of the first to join the team was Scott Miller, a veteran political strategist and adman who, during the years he served as the award-winning creative director at the huge ad agency McCann-Erickson in New York, worked with Miller Brewing, Exxon, and Coca-Cola. He was responsible for Coke's famous "Mean" Joe Greene ads. (A little boy gives Mean Joe a Coke, bringing a rare smile to the notoriously gruff football player's face.) Later, Miller opened his own consultancy where his clients included Apple, Google, Verizon, McDonald's, and Goldman Sachs, as well as politicians such as Boris Yeltsin, Lech Walesa, and George H. W. Bush. Coke's internal motto is to always have its product within "an arm's reach of desire." Miller would also handle the in-store promotions of our effort to make sure that bottled water, too, was within an arm's reach of shoppers.

Our team also included Sergio Fernandez de Cordova, a techno-logical wizard and leading expert on using media to have a social impact. He co-founded a nationwide outdoor advertising company and was recognized as one of the world's leading influencers (that word again).

Sheldon Gilbert came aboard, too. His e-commerce analytics company, Proclivity Systems, deployed patented algorithms to help major retailers predict what consumers will want to buy—even before the consumers know it themselves. He would oversee our online efforts.

## TAKING AIM

To target our messages, we relied on the Natural Marketing Institute's system of "attitudinal segmentation," which divides consumers into five distinct groups depending on how they view food in relation to their health. One segment, called Well Beings, are extremely proac-tive. They fully buy into the health message and eat accordingly. Another group, labeled Food Actives, are concerned about nutrition but not overly passionate. Magic Bullets assume that medicine will fix any health problems their bad diets cause. Fence Sitters really want to do the right thing, but inconvenience and financial obstacles keep them from acting. And as the name implies, Eat, Drink, and Be Merrys don't give a damn. Supersize everything they put into their mouths and they are happy.

We decided that our primary audience should be the Fence Sitters. We wanted to give them one easy, inexpensive action they all could take to become a little bit more healthy. Y&R, the marketing and communications company, created a formal "ad book" like those agencies make up for any campaign—basically a set of messaging guidelines that all of us would follow. The company then developed a series of advertisements featuring historical photographs of Albert Einstein, Muhammad Ali, and Audrey Hepburn drinking water. The message: Water will make you smarter, stronger, and more glamorous. The popular singer/actress Ashanti provided a music video extolling water. It quickly received more than 600,000 social media impressions. The basketball champion Stephen Curry passed up opportunities to endorse soda, instead signing on to a hugely successful campaign with Brita, the water filter company.

Molmol Kuo and her technical team at YesYesNo, a cooperative that aims to put fun into customer experiences, created outdoor drinking fountains for us that "talked." Whenever someone's lips touched the stream of water, thereby activating an electrical circuit, the fountains would say things like "Water keeps your teeth nice and clean. No wonder I was so mesmerized by your smile." "Can I ask you a personal question? Are you a sipper or a gulper?" "When this is all over, I can recommend several excellent public washrooms."

Through a futuristic technique called neuromarketing, where volunteers wear headgear with sensors that measure their brain activity, the external team determined which messages were working and then directed those messages to where our target audience lived and to the media they relied on—television, radio, print, billboards, social media, you name it.

Importantly, we also formed partnerships with more than forty companies, universities, and municipalities to promote Drink Up: convenience store chains, supermarkets, Nestlé. I even managed to arm-twist the soda companies' trade group into supporting us financially by essentially boxing them into a corner. Most big soft drink makers also have bottled water brands. Even though they, too, knew that if their customers drank more water they would drink less soda, they had desperately tried to position themselves as health conscious. I pointed out that they were either with us or against us. If they didn't support us—both with funding and by putting our Drink Up logo on their water bottles—then we could have exposed their hypocrisy. Scott Miller joked that perhaps we couldn't afford the tens of millions of dollars necessary to mount "Cadillac" marketing campaigns like Coke's, but we could sure stick our bumper sticker on Coke's Cadillac and use its products to get our message out.

The final phase of Drink Up was to conduct scientific measurements of the results. That would allow us to find out what was working so that we could build on our success. A division of Nielsen created what researchers call a randomized control trial for us. They found a group of households whose members had never been exposed to Drink Up and another group identical in all ways to the first, except that members had seen our promotions. The results stunned even seasoned marketing executives. For the group familiar with Drink Up messaging, bottled water purchases in stores shot up 3 percent. Orders in restaurants, including fast-food chains, rose by 41 percent.

The most important takeaway is this: If we want to change people's perceptions and choices, we have to speak to their hearts, not their minds, to their aspirations, not just their fears. When we are trying to influence someone's food choices, focusing on emotional connection is essential. In fact, I would go as far as to say the approach by public health advocates, going back decades, has been a failure. Facts and nutritional information alone have not inspired lasting change in eating habits. People are more likely to adjust their food choices when they feel emotionally understood, supported, and engaged. By connecting with their values, preferences, motivations, and aspirations, we can help them align their food choices with what truly matters to them, making those changes more meaningful and long lasting.

Marketing works. If education was the best way to get people to take action or make a different choice, then companies would spend billions on education campaigns. The lesson is applicable to all of us. If you are running a company with healthier or sustainable food, don't only talk about fiber or carbon; connect these benefits to the things people care about. If you work in public health, I would urge you to think about funding campaigns that are rooted in the way people make choices.

And for all the people out there struggling to get their kids, family, or friends to make changes, these lessons are relevant as well. My son Cy loves baseball. When he doesn't want to eat his broccoli, I tell him about the eating habits of his favorite baseball players. Finding what motivates people and connecting that to their choices can help all of us more effectively shape our cultural norms and values. Each time we eat, we are expressing those cultural values, and each time we eat we have a chance to have an impact.

## WITH FRIENDS LIKE THESE . . .

Learning how to most effectively inspire people to make a better choice wasn't the only lesson I learned from Drink Up. We had proven

Watson's concept. Sophisticated marketing could increase demand for healthy products. She became so excited by Drink Up's results that she left her position at Nielsen to start her own communications company that would focus on helping organizations promote good nutrition.

With hard, experimental proof in hand, I was ready to launch other, similar initiatives. If it worked for water, why not use the same template to awaken interest in eating more vegetables, whole-grain bread, or products produced with a lighter carbon footprint?

But something I had not foreseen undermined everything we'd accomplished through Drink Up. What started out as a triumphant victory became a painful defeat. And it came at the hands of the people you'd least expect to sabotage us.

Every day the White House communications office sends out a roundup of recent news stories. All of us (and that included the First Lady) scoured the report, paying close attention to coverage of our own projects. The day after Drink Up was launched, Marion Nestle, the hugely influential Paulette Goddard Professor Emerita of Nutrition, Food Studies, and Public Health at New York University, put up a blog post. I suspected the worst when I read her title: "Drink Up?"

I should say that Marion is a friend and hero of mine. Her work deeply influenced me and shaped my understanding of the gargantuan problems in the food system. But she and other advocates got this one wrong. That snarky little question mark at the end said it all. Before pooh-poohing the core message of the program, she wrote, "Let me be absolutely clear, I'm totally in favor of encouraging kids to drink more water," then she followed that with an emphatic "but" and proceeded into a detailed catalog of the shortcomings of our initiative. Water deficiency is not a health problem in the United States, she wrote. She grumbled that we had paid no attention to the environmental impacts of bottled water. She said drinking water would help combat obesity only if it replaced sugary beverages. She went so far as to claim that soda companies were our main supporters and that failing to use the opportunity to bad-mouth soda was lamentable. Finally: "I'm thinking the White House must have cut a deal with the soda industry along the lines of 'we won't say one word about soda if you will help us promote water, which you bottle lots of under your own brands.' A win-win." (To this day, when I speak in public, I still get people reiterating the narrative that Michelle Obama sold out to industry.)

No, Marion, we did not. That could not have been further from the truth. What we did was launch an effort that could have eliminated tons of sugar from the American diet.

Michael Jacobson, then the executive director of the Center for Science in the Public Interest, long a leading advocate for better food, issued a statement condemning our efforts with faint praise and snide sarcasm: "We're delighted that First Lady Michelle Obama is urging people to drink more water. But we hope people also take that advice to mean that they should drink less soda. There's not exactly a hydration crisis in this country that needs solving."

The two high-profile nutrition experts' messages went viral among progressive organizations and publications like Grist, Civil Eats, and *The Atlantic:* The First Lady has sold out to the soda companies by not telling Americans to drink fewer sugary beverages.

I was floored. Then livid. They had completely missed the point. Before we rolled out Drink Up, we had numerous discussions with food reformers. During these talks, I made two things clear to them. First, I carefully explained to them that Nielsen's research showed irrefutably that when people drink more water, they inevitably drink less soda. I mean, there's only so much room in the human stomach, for God's sake. Second, I said that marketing experts have known for decades that negative health messages simply are far less effective than positive ones. So, why not try an upbeat approach for a change, if only as an experiment to see what the results would be when such a message was applied to promoting public health? No one had done it before. If Drink Up flopped, so be it, but if it succeeded, the potential upside was almost unlimited.

Under Drink Up, soft drink consumption fell without our having to scold the public about its nutritional evils. But those who were supposed to be our supporters reacted to Drink Up with more vehemence than they did several years later when Trump served McDonald's burgers to college football champions in the White House.

Besides, we'd already come out loud and clear telling Americans to drink less soda. We used one of the most prominent soapboxes we had. In 2011, when we unveiled the new MyPlate nutrition graphic, we simplified the hundred-page-plus report produced by the Dietary Guidelines Advisory Committee into two easy-to-read pages focusing on ten important moves anybody could make to improve their health. One of the core official messages was: "Drink water instead of sugary drinks. Cut calories by drinking water or unsweetened beverages. Soda, energy drinks, and sports drinks are a major source of added sugar, and calories, in American diets." None of the previous iterations of the Dietary Guidelines had come down specifically against sugary beverages. Could we have been any clearer?

Before the public relations meltdown, I had hoped that we would be able to build on the momentum of a successful Drink Up campaign. Deploying the enormous powers of modern marketing could have become a template—a model—to show other groups concerned with creating a better food system how they, too, could effectively shift consumers' behavior.

Encouragement from food activists would have helped build that momentum to change our culture. But after the program was called out for being "hypocritical" and in Big Soda's pocket, enthusiasm died. Fundraisers at Partnership for a Healthier America, which we partnered with on the campaign, began to find it hard to get donors willing to provide money to keep the program going. If influential food advocates had rallied behind us, saying that it was a unique marketing effort, the most sophisticated ever launched in the name of better health, money would have flooded into Partnership for a Healthier America.

## INTERNAL BATTLES

The media backlash to Drink Up also dampened the entire Let's Move! effort. The day after we launched the marketing campaign, I was fielding questions from senior staffers and communication leads on what went wrong. How had I let the response be so negative? A few weeks later, I raised the next objective in the strategy at a meeting of East Wing officials. They replied, "So is this going to be a media disaster like the water thing?"

There are internal battles in any company or organization. Everyone has their own priorities, objectives, and agendas. This is especially true in companies, and the people in those settings trying to make their organizations more sustainable or ethical always face an uphill battle. For us in the East Wing, it was a constant tension around which of the First Lady's efforts was going to be the priority. All of us had to maneuver to get more of Michelle. We all pressed to have her do dozens of things at once, far more than she had the time or political capital to undertake. If she was out there as the face of Drink Up, that meant she was not putting her valuable time and image behind a project that someone else in the administration held dear. We were all supposedly in the same boat, but the competition aboard the ship of state was unrelenting, and tempers often fiery.

Although she never rolled in the mud of office politicking, Michelle was fully aware of the value of her reputation, and despite her calm, fun-loving public image she is as savvy and tough as they come. She had extremely high expectations, and there was no space for results that didn't exceed those expectations. She realized all too well that criticism from opponents was all part of trying to get things done in Washington. But it's one thing when Sarah Palin calls you the nanny state as she serves sugar cookies to kids at a rally in Pennsylvania (which she did), and quite another to be criticized by respected authorities who are supposedly on our side. She logically asked herself, "Why the hell am I doing this, when I could be focusing on all these other important things where the people that cared about the issue would actually be supportive?"

Prominent figures had made their reputations as vocal, high-profile critics of the food system. To this day I admire their work. They had played a big role in my understanding of the problems within the food system, and they helped galvanize a generation of people working on these issues. But in the words of Teddy Roosevelt, "Complaining about a problem without proposing a solution is called whining." In their articles and books, these critics offered few ideas or specific plans on how to fix the issues they complained about. Don't get me wrong, calling attention to the problems of Big Food was a critical first step in raising awareness and putting the issues on the cultural map. And critics always have a role in nudging their allies in government forward. No one expected them to be a cheering PR chorus for the administration. On the other hand, no one in a position of power will be able to make the most of their position without advocates who are organized, pragmatic, and strategic. Ones who know how to build momentum, put points on the board, and use that momentum to win the next round, even if they don't get everything they wanted. It is a dynamic that plays out across most areas, and it certainly played out in food.

If administrators in a future White House get real opportunities to improve food, advocates need to realize that supporting their allies in government can at times be far more beneficial than casting aspersions. When we came into power, those hoping for a better food system should have pivoted toward a more strategic approach to change. No longer did stones need to be thrown to get attention; advocates had their own on the inside. Yet instead of backing policies that moved the needle in the right direction on the issues they cared about, they became significant impediments to progress by blunting momentum for the very people pushing the change. At a time when we needed leaders in the food movement, we mostly had critics masquerading as leaders and telling the country that the administration had been co-opted by the food industry. It was a classic case of letting some kind of "purity test" get in the way of actually making concrete progress.

## OUR WAR AGAINST TRANS FAT

Once you have culture securely on your side, there are few limits on how hard you can push for change and have successful results.

I'll break the fourth wall a bit to tell you that this next anecdote is probably more of a fit in the next chapter, "Changing Policy," but I want to show you an example of what's possible when you do successfully move the culture.

The goodwill Michelle generated during the first Obama term clearly bore fruit in 2014 when we began to take on trans fats. For more than two decades, officials inside the U.S. Food and Drug

Administration (FDA) had been trying and failing to advance regulations that would strictly limit, or better yet ban, the use of artificial
trans fats, which were once commonly included in thousands of processed foods to improve stability and texture and extend shelf life.
The same fat could also be reused over and over again in deep fryers
so integral to fast-food kitchens, which meant trans fats were
extremely common in restaurants of all kinds.

Food companies' dream come true was a nightmare for the
nation's health. No respectable nutritionist disputed the science: By
raising so-called bad cholesterol and lowering the "good" kind, trans
fats were considered one of the biggest contributors to fatal heart
attacks, responsible for tens of thousands of deaths. Moreover, the
food industry had less harmful alternatives that worked well.

But because trans fats were cheap and gave products made with
them near-eternal shelf life, the industry clung to them, and the
Grocery Manufacturers Association (GMA), since renamed the
Consumer Brands Association, deployed its money and lobbying
might to stall any efforts to regulate the use of trans fats—health consequences be damned. Earlier administrations buckled, but with the
Obamas in office the long-frustrated FDA officials saw an opportunity to move ahead. A popular First Lady who would have their backs
gave them the political support and cover that they needed to be
aggressive. They also recognized that the country was in the throes of
a cultural transformation. The public liked and trusted Michelle.
Citizens recognized that she had the best interests of kids and families
at heart, and they were beginning to clamor for better food.

That did not prevent the GMA from firing off a final salvo. Louis
Finkel, the organization's government affairs guy, called me on the
telephone declaring that limiting trans fats was governmental overreach. He made the absurd claim that we didn't have scientific backing

for what constituted safe and unsafe levels of trans fats. He threatened that the industry was basically going to go to war against us. He was going to label our efforts the workings of that loathed nanny state.

I let him rant. Finkel was typical of many lobbyists in that he was neither passionate about nor particularly involved in the issues he promoted. He dutifully regurgitated the GMA's official line. To him, it was nothing personal, a job. But it was personal to me. I knew that the culture was clearly moving against food additives like trans fats, and I was certain that it gave me the strength to push back against Finkel—hard.

When he paused in his tirade, I let him have it, profanities and all. "Your fucking industry is putting trans fats in our foods knowing that they are killing people. You've had decades of time to take them out of your products. But if you want to pick a fight with us over an additive that is killing people, well, bring it on, because we will kick your ass up and down the field. I promise, I will come at you as hard as I can."

He sputtered something about how limiting trans fats would deprive millions of voters of such beloved products as Cool Whip.

That made me mad, something I always tried to avoid, even when dealing with the most loathsome of lobbyists. But he was supporting a product that was blatantly undermining the health of Americans.

I said, "Cool Whip! Are you fucking kidding me? First, you can find an alternative to trans fats to use in the stuff, and you damn well know it. Second, if we save tens of thousands of lives and the world no longer has Cool Whip, then we're happy to own that one."

The GMA subsequently made a few feeble efforts to pare back some of the regulations, but it essentially gave up. After the expiration of a three-year grace period, for all practical purposes artificial trans fats disappeared from our tables. That is what can happen when those in government know they have the power of culture behind them.

Toward the end of my White House assignment, I had a parting conversation with Pamela Bailey. She was a consummate Washington insider, sometimes lobbying on behalf of industry, other times working within the government. As the GMA's president at the time, she was Finkel's ultimate boss. Her tenure at the organization began at about the same time as Michelle and I introduced our health initiatives. Bailey told me that when she first arrived on the job, she gathered a large group of leaders from the huge food companies that made up the organization's membership to ask what their priorities for her were. Not one mentioned health, obesity, or diabetes.

Six years passed. American culture changed, and the executives knew it. Sales of processed foods were flattening. In its annual surveys, the International Food Information Council, a group supported by the food and agriculture industries, showed that the number of consumers saying that healthfulness was driving their food-buying decisions had jumped by 22 percent.

Bailey told me that at the most recent gathering of GMA's board, health, obesity, and diabetes were priorities one, two, and three. Then she said something that stunned me. "When I started the job, there wasn't a single CEO that was discussing health and working on what we should do. You and Michelle came in and launched Let's Move!, and now those issues are all the CEOs are talking about. How do we grapple with them? It's where the market is going. It's where the political policy is going. And that is completely because of what Michelle did."

That's not quite true. It was also because millions of shoppers began making the simple decision to put better food in their grocery carts. It sounds like a bromide, I know, but it's critically important to understand the vital role that individual, everyday choices by "normal" shoppers (by that I mean you and me) play. Eaters are not to blame for the situation we find ourselves in; they are a product of the food environment they find themselves in. But eaters have an important role in shifting our culture. So first, fix your own food choices. Make changes yourself, and then don't be hesitant to talk about it. Work to bring the changes you've made yourself to those immediately around you—at your workplace, church, club. Don't pontificate. It's awkward and isn't likely to change anything. But merely asking the right questions can inject climate issues into a group's menu planning and bring them to the attention of leaders. "Can we have more salads at meals? Can we offer alternatives to meat for main courses? Can we buy from nearby farmers?"

The solar industry provides a striking example of the power of personal choice. Did you know that the single most important thing that you can do to encourage the spread of solar panels on homes is to put one on yours? Study after study has shown that the biggest determinant of whether a householder installs solar panels is not government subsidies, not savings on energy bills, not even personal concern for the environment, but whether a neighbor has installed solar panels. Humans are highly social, herd creatures. We take behavioral cues from those around us, and that applies to both solar panels and food.

Profound results can spring from individual decisions, and individual decisions, writ large, are what make our culture.

# Changing Policy

Idealism is well and good—even necessary—but too much of it can stifle change for the better. Early on, I was as idealistic as the next conscientious eater, although it took some time for me to get there.

At the beginning of my professional cooking career, I too had a hyper-focused view of the role of a chef. At age twenty-two, my total experience was a summer job at an Italian place in Chicago that specialized in serving lunch to nearby office workers and feeding the pretheater crowd. But I lucked into an unsalaried apprenticeship at a Michelin-starred restaurant in Vienna, Austria, where I was spending my last quarter at the University of Chicago in a study abroad program. Determined to prove myself, and to learn as much as I could (since knowledge was my pay), I threw myself into the challenge of becoming the best damn line cook anywhere. For more than a year, my focus stayed on my cutting board and the saucepans on the burners in front of me, except during my off-hours, when, exhausted, I often fell asleep studying cookbooks by Charlie Trotter, Thomas Keller, Paul Bocuse, and other legendary chefs, hoping to absorb their secrets.

And I did get better, but far from perfect. One afternoon, the sous-chef Alois, who was my mentor, asked me to create a rhubarb sauce to accompany seared foie gras. "Yankee," he said. "Cook the rhubarb down and fuck in the butter," betraying a less-than-fluent mastery of that multipurpose English vulgarism.

Nevertheless, I understood what he meant and plunked a fist-sized lump of butter into the saucepan. He gave me a withering look. "I said *fuck* in the butter." Dutifully I added another enormous hunk. I mean, this was an accompaniment to fatty goose liver, for God's sake. Perhaps sensing my hesitation, he came over to my station and got right up close to my face and made a pronouncement that forever changed my approach to professional cooking, though not in the way he intended. "If the guest walks out of the restaurant and drops dead of a heart attack, it's not my problem," he said. "The guests are asking us to make food that tastes good, not what's good for them." He leaned in front of me and scooped another grapefruit-sized poultice of butter and plopped it into the pan.

I went back to my work, embarrassed. Alois remains the single best chef I've ever worked with. He was a machine in the kitchen. Everything he did was precise, and he strove for perfection in every single thing he made, and he almost always achieved it. His station was always immaculate, his white chef coat spotless at all times. He never took shortcuts. After service he (we) would stay out all night drinking, and in the morning he would ride his racing bicycle for hours at a relentless pace. I'll forever respect his professionalism and skill, but in that moment my mind began to race.

On the one hand, he was entirely right, that is exactly what we as eaters were asking of chefs . . . and of food companies. At the same time, I felt there was something fundamentally wrong with this. It sounded totally screwed up. Looking up and glancing directly to the right from the station where I had stood for hundreds of nights, I could see into the dining room, with its well-to-do clientele and their double chins, button-straining bellies, and red complexions of people primed for cardiac arrest or stroke. As they all stuffed their faces ravenously with food I had prepared, I asked myself for the first time about the implications of my cooking on the health of the people I was feeding.

While I was absorbed in those thoughts, a door immediately to the left of my station burst open. The purveyor of our chickens, ducks, and eggs entered from the back door, wheeling a dolly laden with boxes that he left in a stack for us to unpack. His timely entrance prompted me to ask myself, "And what is the impact of the food I cooked on the growers, the land, the environment?" The kitchen layout brought these issues to light for me starkly. As a chef I was literally and figuratively standing between the people who produce our food and the people who eat it.

It was a set of questions that permanently changed my career. What I did in the kitchen not only affected the health of diners; it also affected the hardworking farmers, fishers, and foragers who wheeled their products through the kitchen door. And their actions in turn profoundly affected the soil, the water, and the environment as a whole, something I didn't fully appreciate until years later when we all became more aware of climate change and food production's role in making it more severe. As a cook, I stood at the center.

I stopped perusing books by Thomas Keller and began reading Michael Pollan, Marion Nestle, and every book I could get my hands on about the history of agriculture and food policy. I shopped at farmers markets. I ate organic. My beef was grass fed. I thought that everyone should eat that way. When I arrived in Washington, I came with a copy of Michael Pollan's *Omnivore's Dilemma* and was ready to decisively take on Big Ag—until reality reared its ugly head.

Pollan, in his compelling prose and description of an elegant, if harmful, system of corporate-controlled agriculture, had convinced me that if we could cut direct subsidies to growers of commodity crops such as corn and soybeans, the cost of junk food, including cheap fast-food burgers, would soar, making fruits and vegetables more reasonably priced in comparison. Americans would eat more nutritious foods, and the environment would be saved, if we could only get rid of those corn and soy subsidies. This simple thesis remains pervasive today in the minds of most who are engaged with this issue.

It sounded good to me, too, but before I launched into an antisubsidy campaign, I thought I needed to educate myself about how government grants to large commodity farmers actually worked—knowledge that would be the first step, I figured, in eliminating them. It soon became apparent that the issue was far more complex than that neat storyline presented by Pollan and many others in the good-food movement.

It is true that in the 1970s, policy levers—including subsidies—transformed what we grow and how we grow it. Under Secretary of Agriculture Earl Butz, appointed by President Nixon, U.S. farm policy shifted from supply management (paying farmers to let land lie fallow to avoid overproduction) to a policy that encouraged maximum production—summed up in his famous phrase: "Get big or get out." This change promoted corn and soybean production through subsidies and guaranteed prices, creating a lasting agricultural system focused on a few staple crops; this devastated smaller family farms.

But what was the current impact of the subsidies? If we eliminated them, what would be the impact on the relative cost of healthy food versus hyperprocessed food? How many acres of agricultural land could we expect to shift if we were successful?

I talked to every agricultural economist I could. These folks were numbers people; they would let the math speak for itself and leave the debate on the policy up to the politicians and political appointees. I found only *one* guy at the USDA who estimated that removing subsidies from corn and soy would actually make a difference. And *his* predictions showed that if we got rid of those direct subsidies, we would shift those acres of soy to other uses by 1 percent. I had no idea from reading Pollan that subsidies amounted to such a small fraction of the value of the corn crop. In 2016, for instance, U.S. corn farmers received less than $2.2 billion in government subsidies in the form of insurance, which is a lot of money that could do real good elsewhere, no doubt, but it was chump change to the industry in a year when they sold *$50.4 billion worth of corn.* Those subsidies amounted to the value of 4.4 percent of total production. For soy, in 2017, growers produced $41.3 billion worth and received $1.6 billion, an even smaller percentage. In the overall picture of the ag economy, subsidies were not the reason growers were growing what they grew.

In truth, if we simply pulled the cord on subsidies, what we grow wouldn't look fundamentally different than it does today. It is not the panacea I was led to believe.

But those subsidies were a very attractive red herring. I get it. We all want to know that there is a silver bullet to solve all our problems somewhere. And on some very deep, human level, it's focusing and oddly comforting to believe that silver bullet was attainable if only you could get the right people to see your perfect logic.

The less-than-idealistic and less-than-simple reality, I discovered, was that a fast-food cheeseburger is cheap because decades of investments and innovation have led to advances in seed genetics, fertilizers, pesticides, and farm machinery. Technology has allowed U.S. corn and soy farmers to become extremely efficient (if you ignore the externalized costs of the environmental damage they wreak). In the 1920s and 1930s the average acre produced between 20 and 30 bushels of corn per year. By 1960 the average jumped to about 50 to 60 bushels per acre per year. Today, the average acre in the United States is yielding between 170 and 180 per acre. A typical American farmer in the 1940s might have harvested 100 bushels of corn in a nine-hour workday. A farmer aboard a modern combine can bring in more than 8,000 bushels. This efficiency and massive scale is by far the greatest driver for cheap food, far more than support from the government.

On the other hand, growers of produce are still reliant on much more manual labor, just as they have always been. Most fruit and vegetable production still requires substantial labor, which keeps the costs of these foods elevated. There are also other factors—like cold supply chains and spoilage for fruits and vegetables—that contribute to the relative price difference. Many point out, "Why don't we take the subsidies from the food that makes us unhealthy and subsidize foods that are healthy?" That is a fair question. Ironically, the people who currently grow fruits and vegetables are the ones in D.C. fighting against that. Politicians from major fruit- and vegetable-growing states, such as the former House Speaker Nancy Pelosi, a Californian and hero of mine, blocked support for subsidies for produce farmers. That's because the status quo enables their constituents to use advantages such as ideal weather and rich soil to gain a competitive edge over farmers in other regions and keep new entrants out of the market. In Washington, little progress is possible when the strategy boils down to trying to take money from extremely powerful groups (big commodity ag) who have widely distributed power and give it to much smaller groups (specialty producers) who fight against that very change.

Changing the subsidy system is important, but will be inherently slow and incremental and is not the magic wand we have been led to believe. In addition to the focus on trying to change subsidies, there are a multitude of policy approaches at all levels of government that we should pursue. We should be investing in the research and development of the same sorts of innovations for produce growers that farmers who raise commodity grains enjoy. That would enable growers of more nutrient-dense food to benefit from similar reductions in the cost of materials and labor while increasing yields. We need to reduce the unforgivable amount of food waste of produce farming by supporting organizations that glean fields, and supermarkets that donate slightly dated food to charities. We should perhaps consider buying less-than-perfect fruits and vegetables ourselves. Nationally changing sell-by date policy would go a long way to reducing food loss and waste. Don't get me wrong: I believe the current subsidy system—which now mostly resides in insurance policy after we eliminated direct payments—is a poor use of taxpayer money. That money would be much better spent incentivizing a transition to regenerative agricultural systems or supporting better fruit and vegetable production. But I don't know anyone in Washington who thinks that we will eliminate corn and soy subsidies in our lifetime. Still, I do think we can find ways to add more provisions over time to incentivize practices and behaviors that are better for the planet when it comes to these large commodity programs.

## ALL POLITICS IS LOCAL

Although it might not be as sexy or exciting as fighting the good fight at the national level, there is no more important work than the efforts to create grassroots changes at local and state levels. This is how our system was designed. Allow states and local governments to experiment and enact policies that meet the needs of their communities. By nature timid when facing voters and entrenched business interests, Washington policy makers abhor experimentation. When policies affect tens if not hundreds of millions of people, the national level is not usually the place to roll the dice and try something out. Federal policy makers are much more amenable to taking the best tried-and-true local approaches to dealing with issues and bringing them up to a national scale. Before making a decision on a national policy in the White House, we would always have to know what states had tried it and what the results were. One example of this was when we were trying to finalize the menu labeling rule that was part of the Affordable Care Act.

When we were working to formulate nationwide regulations requiring restaurants and other food outlets to include calorie counts on their menus, we ran into tremendous resistance from the movie theater industry. It might not seem obvious when you queue for the latest blockbuster, but cinemas are actually as much in the business of selling junk food as they are in the business of selling tickets. Their concession stands generate profit margins of 85 percent (thanks to the seven bucks you part with for a medium soft drink). Ticket sales generate less than half that margin. With buckets of popcorn delivering more than a thousand calories each (Popcorn? Who knew?), it was easy to understand why the National Association of Theatre Owners, a lobby group, went to war with those of us who said their concession stands should be covered by calorie-count regulations just like restaurant chains, grocery stores, and pizza delivery companies.

Denis McDonough, then Barack's chief of staff, had run the National Security Council for the president in his previous role. A devout Catholic and avid runner, Denis is a kind, principled, decent man who always tried to do what was right. He was also tough as shit and someone you were glad to have on your side. In a meeting in his office, he wondered if including theaters was worth the political capital—especially because these discussions were taking place just before the 2012 election and we didn't want to turn off voters. His first question to us was, "Has any other jurisdiction attempted something like this?" His second: "If so, what happened?"

Those of us pushing for theaters to be covered by the rule had done our homework on this question. Thanks to local, grassroots efforts, both New York City and the State of California had already implemented laws requiring menu labeling for theaters. We knew how they had crafted their regulations and where they drew the lines between those businesses that would have to comply and those given a pass. We checked with advocacy groups and found from them what, if any, political opposition had resulted. Ultimately, we assured McDonough that the sky had not fallen. In fact, nothing much happened at all. Given the all clear, we included theaters under the federal labeling regulations. The local efforts cleared our path. If we had come back and told him that menu labeling had never been tried before, our discussions would have ended abruptly with McDonough.

So what does this mean for you? The truth is that it's never enough to vote only once every four years when there's a presidential race. You've got to stay engaged. Vote not just every four years, and not just every two when Congress is up for election, but every year, when local races are determined. Local policy makers are the ones who create models for Washington to take up. You've heard of the phrase that the states are the "laboratories of democracy," right? Nothing could be more true. And what's more, being engaged in local politics also means that you may have more of a say than you realize. Local leaders and their staffs can be surprisingly receptive to their constituents' calls, letters, and meetings. They're your neighbors, they shop at the same stores, they go to your same houses of worship. Real change always starts from the ground up.

## KIDS' FOOD

Our success in getting the Healthy, Hunger-Free Kids Act passed serves as an example of how advocates, administrators, and politicians should approach fixing food policy. Early in the administration, the Child Nutrition Act of 1966 was due for reauthorization. That act, which is reviewed every five years, is known to most Americans through the school lunch program. It had previously been rubber-stamped with few, if any, changes. Melody Barnes and Martha Coven of the Domestic Policy Council, which advises the president on domestic issues, told us that despite its sleepy reputation, the act offered numerous opportunities to improve the food children received at school. Another benefit was that reauthorization had to happen in 2010, giving us (and Congress) a deadline. So, fixing the Child Nutrition Act became our top priority.

We began cautiously. Having seen the debacle that resulted from Hillary Clinton's health-care reform efforts during her husband's first term, the last thing we wanted was for Michelle to *appear* heavy-handed in matters beyond the accepted purview of a First Lady in ways that would undermine the chances of getting the bill passed. Notice that I italicized the word "appear." Behind the scenes and out of publicity's glare, she would be putting all of her effort and influence into pushing the act—with the changes she wanted—through Congress.

We started off with a series of internal meetings with officials from the Departments of Health and Human Services and Agriculture. Our discussions expanded to include the Centers for Disease Control and Prevention and the military leadership who were concerned about the high percentage of potential recruits who failed to meet basic health and fitness criteria for enlistment.

Eventually we reached out to nongovernmental groups such as the Center for Science in the Public Interest, the American Heart Association, the Kellogg Foundation (which works to improve the lot of children in vulnerable communities), the National Education Association, the National Association of Educators, the National School Boards Association, the School Nutrition Association, Feeding America (a charity that fights hunger), cafeteria workers organizations, food companies, and several other groups that had an interest in better nutrition for schoolchildren.

These meetings had two important results. One, there was a broad spectrum of allies who were all interested in improving the badly outdated Child Nutrition Act. Two, partners supplied us with a wealth of valuable ideas on what should go into the renewed act.

Some of our intentions we deliberately wouldn't mention in speeches and at events to avoid rousing conservatives' ire. A particularly sensitive provision was a plan that came to be called the Community Eligibility Program. It would fund schools in poorer districts to feed both breakfast and lunch to *all* their students, at no cost to the kids. Normally, students would be divided up: Kids below the poverty line would get free lunch and kids close to the poverty line would get a reduced-cost lunch, while kids from families of greater means paid full price. But at lunch, everyone ate together. No one could tell the difference between who was on free lunch and who wasn't.

The same hadn't been true for breakfast. Breakfast programs were only for poor kids. Because of the stigma of being poor, kids across the country would forgo breakfast even though many of them hadn't eaten since school lunch the day before, because they didn't want to feel the shame associated with not being able to afford food. Just think about that. Think about how powerful those feelings are for an eight-year-old or a twelve-year-old. And think about the damage caused by those kids trying to learn with stomach pains from hunger. The provision provided breakfast to all kids in schools that had more than 40 percent of kids who were low income. The change to feed everyone was a powerful policy for the poorest kids. No longer would students have to suffer the shame of getting free food in front of classmates who came to school with their bellies full. Everyone attending the schools in these districts, food secure and food insecure alike, would be able to have the free meals. In essence, we were creating a vast, national antihunger and antipoverty program, although we never advertised it as such. For what to the federal government is a minimal cost, we could not only head off hunger and shame but also see the improved academic outcomes of students with fuel in their bodies and brains—a well-studied and established fact.

Once we had a draft of the bill, we embarked on the policy equivalent of a full-court press. The White House Office of Legislative Affairs began reaching out to senators and members of Congress to promote the bill. The outside groups that supported us activated their own local memberships to lobby representatives. The School Nutrition Association flew "lunch ladies" (or school chefs, as I prefer to call them) from across the country into Washington to meet with legislators from their home districts. And then we had Michelle, who promoted the bill for months on end. After one particularly frustrating day, she sighed and said to me, "They need to move this bill. I am done with this. I am tired of saying the same thing over and over when this is obviously what is best for our kids. This is ridiculous."

But thankfully, she kept talking, and she kept pushing.

Michelle aimed her message at parents who try to feed their kids healthy food at home, only to have their efforts thwarted by school cafeterias and the junk-filled vending machines in school hallways. She took the position that tax dollars should not be spent to nullify the good intentions of parents. Polls showed that an astounding 90 percent of voters wanted better school lunches for their children. All of this represented a powerful force that put enormous pressure on politicians. Even so, with the closing days of the 111th Congress approaching, getting the act through in time looked increasingly doubtful, especially in view of the legislature's packed calendar.

At the time, Obama and Democrats in the Senate and House were focused on trying to reverse the worst economic catastrophe since the

Great Depression. They did this against a backdrop of Republican vitriol and negativism. Legislators were struggling to pass the Dodd-Frank Wall Street Reform and Consumer Protection Act, the Job Creation Act, and other bills intended to keep the country from a complete economic collapse. It is hard to convey the weight of the pressure that the president and his team were under. There was no time and very little urgency for school lunches. The Child Nutrition bill stalled, and stayed that way for months. Even after the Senate passed the act unanimously, the House continued to sit on it.

Michelle grew frustrated. She began personally calling members of Congress, something the vast majority of First Ladies had never done. She made it clear to her husband that even in such hard times healthy kids were a priority for the nation's long-term success.

Getting the law passed immediately was as vital as anything else Congress was dealing with. Only a few weeks remained before a new session of Congress would begin—one with Republicans in the majority. Under their leadership, the bill would not have had a chance of getting enacted.

As the ultimate source of pressure, the president approached the then House minority whip, Eric Cantor, who had been using procedural rules to hold up the legislation. As hard as it is to envision, especially given the state of relations between Republican and Democratic politicians today, give-and-take was still the currency upon which legislation functioned. Obama made it clear to Cantor that the law was of extreme importance to him—and his administration, first among them the First Lady. Letting it advance to a vote required that Cantor do absolutely nothing other than step out of the way. He could give the president a big win and chalk up a mountain of political goodwill at little cost to himself. Half jokingly, Barack closed the deal by saying, "I need this to get done; otherwise, I'm going to be sleeping on the couch."

The bill passed just weeks before the session ended.

Traditionally, a president holds signing ceremonies for important laws in the Oval Office, with him sitting majestically behind the Resolute desk encircled by a small group of witnesses. For the Healthy, Hunger-Free Kids Act, several hundred of us—politicians, administrators, teachers, and students—gathered in the gym of Tubman Elementary School in Washington, D.C. During a brief address, Barack admitted that the bill would never have passed without Michelle's pushing for it and alluded to his plea to Cantor about being relegated to the couch if it hadn't.

"We won't go into that," Michelle interjected with perfect comedic timing. "Let's say it got done, so we don't have to go down that road."

I was standing in the back of the room. While Michelle spoke, Barack caught my eye and shot me a quick wink. It is hard to describe how that moment felt. That wink was an acknowledgment of the unlikely journey we were on, starting from their kitchen in Chicago. That we had just accomplished something truly significant, helping millions of kids every day who needed it most. And he knew that I knew how it really happened. It was an amazing feeling. Our bill had passed with bipartisan support in Congress, that rarest of rare achievements in the Obama era. We not only had passed the first increase in federal funding for school nutrition programs in fifteen years but, as the president pointed out, had done so in a way that did not add a dime to the federal deficit.

Nutrition standards that had remained unchanged since before anyone had even thought that an epidemic of childhood obesity was possible were brought in line with the latest science-based recommendations from the National Academy of Medicine. For fifty million schoolkids, levels of sugar, fat, and salt went down. Menus had to offer more fruits, non-starchy vegetables, and whole grains, which, in fact, completely replaced ultrarefined, processed grains in lunches.

Vending machines could no longer dispense candy and sugary soda, items that were formerly the only things some students consumed during the day.

As he signed, I realized that more than thirty million kids, most of them poor, would be benefiting in a profound way from what was happening. Not only would they be getting free breakfasts and lunches, but those meals would be much more nutritious than what was previously on offer.

And our efforts have been vindicated. In 2020, Harvard University researchers released the results of a survey looking into the effects of the Healthy, Hunger-Free Kids Act. For poorer children, whose rates of obesity had been growing at an alarmingly fast rate before the changes, the risk of their being obese was nearly 50 percent lower than it would have been on the earlier trajectory.

The Community Eligibility Provision deserves a lot of credit. It allowed the government to get healthy food to the kids who needed it most, especially breakfast, a meal unavailable to many at home. In 2010, just over 11 million kids participated in the school breakfast; as of 2022, 15.7 million kids were being supported by the program, and 87 percent of them were low income. Just as important, younger students saw for the first time what a plate of nutritious food actually looked like. This has the potential to be a new normal that they will carry through life.

For food activists, there are two lessons to be taken from the Healthy, Hunger-Free Kids Act's success. First, it serves as an example of how good policy gets made. The coalition of advocates had done a fantastic job. They were pragmatic, organized, relentless, and collaborative. We simply would not have gotten this bill passed if it weren't for their effectiveness on the Hill and in the public. Second, because we got it through by the skin of our teeth, it's a lesson in how things cannot get done in Washington without the groundwork of shaping the culture, gaining a broad coalition of supporters, crafting sound policies, and having leadership from the White House. Passing federal laws is *meant* to be hard, for sound reasons. It should be tough to do things that affect every resident of the country. So, that means we better have the will to do the work.

## FOOD MOVEMENT, WHAT FOOD MOVEMENT?

During our time in office there was a lot of talk of the food movement. For years people had been working to change food in the country, and change was happening little by little. When Michelle started digging up the South Grounds for the garden and launched all of her work on improving food, it felt as if the "movement" had arrived. Good food was in the White House, fighting for change.

On some level that sentiment was right. Good food was in the White House, and if it wasn't for the work of people over generations from every corner of the food world, we never would have focused on these issues. We were a result of an initial cultural shift. But to truly be an effective influence on government policy, we need to take an unbiased look at ourselves and be honest about where this work stands. An unbiased view would have shown that a cultural shift was under way back then that still continues today, but the cultural shift has not yet translated into a true movement. Michelle did not take on the issues of food because voters were calling on the administration to do so. Advocates did not use their pulpits to persuade voters to support Barack Obama or a significant number of state or national legislators (if any at all) because of the food policies they would champion. As a result, none of those elected officials had a sense of accountability to voters who truly cared about food issues. Michelle chose to take on nutrition because she was a mom, not a politician.

She knew how tough it was for a working mother to make sure her kids ate well. She felt that using her position to improve food could have a profound impact. It was something that she could use the Office of the First Lady to successfully champion.

Conscientious eaters—and I include myself—have lived within our own feedback loop: an insular, like-minded community whose members are super-aware of what they eat and its effects on the environment and human health. This causes many of us to suffer from a level of delusion that the movement is a lot further along and more powerful than it is. In the early years of this century, people who championed a better food system often considered themselves part of a movement making organic, local, and unprocessed food an unstoppable wave of the future. In truth, in the political, policy-influencing sense, it wasn't a true movement at all, as Barack once explained to Dan Barber, one of the most well-known and conscientious chefs in the country.

Barber came to Chicago between the 2008 election and the inauguration to cook dinner at the Obamas' house, at Barack and Michelle's invitation. The Obamas had asked me what chef I admired the most and whom I would be most excited to potentially work with at the White House. I cold-called Dan Barber. After a couple funny conversations where he thought I was pranking him—"Are you fucking with me, because I don't have time for this if what you are saying is bullshit"—and an impromptu meeting in Chicago, he realized I wasn't lying to him and the invitation to come cook dinner with me was real.

About a month after President-elect Obama won the election, the renowned chef arrived at the back door on a bitterly cold late-fall day with two enormous coolers, having brought every ingredient needed for dinner with him from the farm surrounding his restaurant, Blue Hill at Stone Barns, just north of New York City. He had even transported the cooking and eating utensils and dishware we'd be using.

The meal was delicious: scallops, chicken wings sous vide, and—in a bold move—a rich, intensely flavorful braised lamb neck. Afterward, the Obamas joined Barber and me in the kitchen while we cleaned up. Barack asked Dan about his thoughts on food production, and Barber began talking about the ecological evils of mono-cropping corn and soybeans and the need to shift over to diverse, nutrient-dense crops.

"I totally agree with you," Barack said.

But as a senator from a farm state like Illinois, he knew firsthand how big, powerful, and deeply embedded in politics industrial-scale agriculture is. "Taking it on is very hard," he said. "And if I try to head into the inevitable storm without the support of a true movement, I'm not going to get very far." He stopped, looked hard at Barber, and said, "Make me do it. Show me the movement."

You only have to consider what has happened to it in the years since to see how fledgling the "food movement" truly was. With the exception of some antihunger efforts and help from an environmental movement that has only more recently taken an interest in food production's contribution to climate change, food activists on a national stage have largely gone quiet and all but disappeared from the Washington radar, beyond a few dedicated groups doing hand-to-hand combat on Capitol Hill, trying to defend progressive policies and make small gains where they can.

To make any policy headway in Washington, a robust coalition of organizations needs to be generally unified and ready with a detailed agenda and well-thought-through proposals. They also need a coordinated strategy on where they are headed and the means they intend to use to get there. Our opponents in the food industry certainly have one.

As David Plouffe, Obama's campaign manager, said in his book *The Audacity to Win*, "In any organization, you have to determine your pathway to success and commit to it. There will inevitably be highs and lows. But you have to give your theory and strategy time to work. Maybe it won't. Maybe endeavors fail. But without a clear sense of where you are headed, you will almost certainly fail."

So, what would a real, effective food movement look like? The environmental movement is a good model. I recall the glaring difference in meetings with foodies and with environmental groups, where a small army of experts arrived at the White House. Environmentalists walked in with teams of seasoned lobbyists and lawyers with far more experience in Washington than many of us on staff. Representatives of several environmental organizations often joined forces to press for mutually important policies. They would present us with polling and survey data favorable to their position. They had fleshed out legislative and policy proposals good enough to be used off the shelf with minimal tweaking on our part. Not only were their D.C. policy machines well run, well organized, and well funded, but they had offices working at state and local levels to help them put grassroots pressure on senators and members of Congress.

Even then, success was hard and took decades. The Inflation Reduction Act, signed into law by President Biden, which will go down as maybe the most important piece of climate legislation anywhere on the globe, was the result of decades of work, organizing and building a real environmental movement.

With food advocates, we never experienced anything close to this level of sophistication outside the Child Nutrition Act. Don't get me wrong, there are some great organizations doing great work. A few organizations such as the Center for Science in the Public Interest, the American Heart Association, the Environmental Working Group, and some antihunger outfits like the Food Research and Action Center were effective, but at best food had only the bare bones of a real presence in Washington. Much work remains to build the broad support, resources, and infrastructure to meaningfully accelerate the progress from policy leaders.

That is starting to change. The infrastructure of a movement is being built. When we got into office, New York University was the only college that offered a master's degree in food studies outside agricultural programs at land grant universities. Now there are dozens of programs across the country. Numerous law schools offer programs to train future lawyers around food policy and law. Food policy councils are becoming commonplace in state and local governments. And there is a lot of progress being made to integrate the food and climate movements.

In the second term of the Obama administration, it had become clear to me that those of us in the good-food world would do better by working to integrate our priorities into broader coalitions, as opposed to trying to create a stand-alone movement. The climate movement had begun to gain traction with the culmination of a global agreement to combat climate change coming out of COP 21 in Paris (yes, the conference where my Last Supper event nearly flew off the rails). The best way forward for food activists might be to join the climate change movement—a big ship with plenty of room aboard. It has infrastructure that food advocates lacked. Many groups had come together under the climate banner. They had attracted plenty of funding. Some of the smartest scientists in the world were figuring out what needed to be done and how to make sure it got done. It was nascent at best back then, but the issues of food production and environmental degradation are becoming increasingly glaring, and focus on food and agriculture is finally on the rise.

Groups like the Nature Conservancy, Conservation International, the World Wildlife Fund, the World Resources Institute, the National Resources Defense Council, and the Environmental Defense Fund, to name a few, have all increased their focus on the food system over the past fifteen years. While most of the environmental movement and flows of capital in both the public and the private sectors are still overwhelmingly focused on energy transition, the food industry increasingly needs to be center stage. The more collaboration between groups working on climate and those working on food, the further we will get. Food is the language that can explain and connect the implications of climate change to the general public. And if we don't solve food, none of the efforts toward renewable energy or electric cars will make much difference. We simply won't solve climate if we don't solve food.

## SO WHAT CAN WE EXPECT FROM POLICY?

The short answer is that I do not think that we can view policy change as a silver bullet that can fix our food system. Especially so if we haven't yet shifted the culture in our favor. But make no mistake: It's worth putting effort and long-term strategy into policy solutions; we won't accomplish the change we need if we don't shift policy.

Believe it or not, the Trump administration has potentially turned the political dynamic of food policy issues on its head, with the appointment of Robert F. Kennedy Jr. to lead the Department of Health and Human Services (HHS). The Make America Healthy Again (MAHA) slogan has some elements of a movement that, frankly, I would have *loved* to have when we were in the White House: MAHA has a number of influential voices on social media with substantial followings getting messages out and galvanizing stakeholder engagement. More important, Trump and RFK Jr. have—for the time being, as I write this in the spring of 2025—made the issue of health and food bipartisan. I continue to be stunned to see Republicans and Fox News hosts ring the alarm bell on health (though not climate, of course) and start pushing legislation to try to improve the system.

These are the same people who attacked us at every turn and decried our "nanny state overreach." I am skeptical that this bipartisanship will last, but it would be transformational for our issues if it remained.

RFK Jr.'s critique of the food system—that it prioritizes profit over nutrition and real food—is one that I, and most of us in the food world, have heard for some time and largely agree with. He articulates it with force, and for those of us who care about these issues, it's evocative to hear someone in a position of real power give these concerns voice. While it is impossible to render a final judgment—at the time of the final edit of this book, we were only a few months into his stint in government—it does seem that the major benefit of this unholy alliance ends there. Under Trump, RFK Jr. has already disrupted the food-policy landscape but is doing so by riddling it with conspiracy theories, bunk science, misplaced priorities, and downright dangerous policies. He has overseen the gutting of HHS staff. HHS oversees the Food and Drug Administration and is the very institution we will need to regulate the food industry more forcefully. He is very visibly prioritizing issues that essentially have no basis in science, such as trying to replace seed oils for beef tallow. He has gone so far as to do a live interview at a fast-food restaurant, Steak 'n Shake, touting a cheeseburger with fries as "healthy" because they switched to using tallow. If the country continues to eat oversized burgers and fries as our daily staple, we will continue to be sick regardless of what oil they're cooked in. RFK Jr.'s conspiracy-theory-based attacks on vaccines are already leading to deaths in Texas from measles outbreaks. What's worse is that he has given this administration a kind of legitimacy in the food world for talking the talk of "making America healthy again," while Elon Musk's DOGE and other forces in the administration are gutting the dollars and policies around climate change and environmental protection, severely hindering our ability to grow food well into the future. They are dismantling the departments we so desperately need to help us meet the stark challenges ahead.

All that said, there's no way forward but to move forward. While this administration has shown itself to be hostile toward many of the policies I'll outline below, it's vital to remember that policymaking is by nature a long-term game. The need for better policy will outlive this administration. True, perhaps the timeline will be different and new facts on the ground will cause us to have to change our strategies, but politics and good policy advocacy cannot stop even when we have a hostile president in office.

Policy is a critical component of change. It is particularly important to set a baseline of behavior in the industry, ensuring that the worst practices, which harm both people and the planet, are prevented. These include basic protections such as food safety regulations. Strong policies can prevent incidents like the listeria outbreak at the Boar's Head meat-processing plant in 2024, which killed ten people and sickened dozens more. This tragedy followed a rollback in safety standards for meat slaughter and processing plants under the Trump administration.

There are countless areas where we need strong policy to protect the public from harm. It doesn't make for the most enthralling reading to go through a comprehensive list of policy proposals, but here are a few examples that would make a real difference:

- Revising the "generally recognized as safe" (GRAS) policy, which currently allows thousands of additives to be used in food without sufficient safety testing.
- Cracking down on manipulative labeling, such as using the terms "natural" and "healthy," which mislead consumers into believing they're making healthier choices when they may not be.

- Continually improving nutritional standards for school and military food programs.
- Reforming sell-by date policies to reduce food waste; in an environment where food waste is a *massive* contributor to greenhouse gas production and climate change, we want to encourage people to eat food that is still perfectly safe. (Most sell-by dates actually have nothing to do with the safety of the food.)
- Enhancing worker protections by ensuring fair pay and better working conditions.

Policy should not only set a baseline to prevent harmful practices but also play a proactive role in creating incentives and support systems that drive the food industry in a better direction. In many ways, the government is better positioned to use policy to protect our environment by incentivizing and de-risking a transition to a more regenerative food system. Future policy efforts should prioritize reducing greenhouse gas emissions, improving soil health, protecting water resources, enhancing biodiversity, and promoting sustainable farming practices.

## Other Key Policy Areas to Prioritize for Climate Resilience

- Sustainable Land Use and Soil Health: Policies that promote regenerative farming practices are essential for restoring soil health, which impacts climate and the nutrient density of food.
- Reducing Agricultural Emissions: Methane and nitrous oxide emissions from agriculture must be addressed

through policies that encourage methane capture, better fertilizer management, and carbon farming.

- Water Conservation: Given increasing water scarcity, policies should encourage water-efficient farming methods and drought-resistant crops.
- Support for Small Farmers: Policies that support small and medium-sized farmers are crucial for promoting sustainable farming practices and ensuring a diverse food system. The loss of small and midsize farms is one of the greatest threats we face in building a resilient food system. The average age of a farmer in the United States is fifty-eight years old, which means we face another massive round of agricultural consolidation in the coming decade. Ensuring we have a diverse base of farmers is essential to manage through the increased volatility we face today and in the future.

The Biden administration, under the leadership of Secretary Vilsack and Robert Bonnie, made tremendous progress in many of these areas. They laid a strong foundation, but much work remains to ensure that most U.S. farmland is cultivated in a way that ensures that future generations will enjoy the same bounty we do. The Trump administration is hard at work trying to undo much of this progress, but these programs were widely popular with growers around the country, so I am hopeful that they may be able to prevent a full-scale rollback.

One of the next key steps the government can take is to establish a carbon market and broader ecosystem services market, enabling farmers to be paid for their role in helping to solve the most existential threat humanity has faced. To achieve producers' transition at scale, we have to make it economically rational for them to do so.

Helping to create the conditions and infrastructure to enable this—rewarding growers for the outcomes they create—is a perfect application of smart policy. While this seems like a distant fantasy in the current political climate, implementing a carbon tax and cap-and-trade system will be the essential unlock to decarbonization in agriculture and the broader economy.

Beyond carbon, conservation programs currently implemented by the USDA—such as the Conservation Stewardship Program, Conservation Reserve Program, and Environmental Quality Incentives Program—are highly effective and must be expanded. Ultimately, elements of these conservation programs should be embedded into any government subsidy, including crop insurance, so we can protect our natural resources and sensitive ecosystems. Insurance needs to be restructured to reward environmentally friendly practices by making them a condition for receiving insurance.

## Climate Adaptation and Disaster Preparedness

The one area that has received woefully little attention and funding is adaptation and reliance. There has long been hesitation in the environmental community to talk openly about resilience and adaptation. The fear is that if we shift our focus, then we are essentially going to give up the fight to decarbonize to stay within a level of warming that would prevent calamity. From the food and agriculture standpoint, the evidence is clear that we are far beyond the point where we can postpone serious investment and effort to ensuring we can produce enough high-quality food at prices people can afford in the decades ahead. We are at 1.5 degrees Celsius of warming now, and we almost certainly will go past two degrees in the coming years. This was true even before that large-scale pullback of climate policy in the United States and increasingly in the EU.

We need to be honest with ourselves about where we are and grapple with the implications of that reality. We are simply unprepared to produce a stable supply of food in a highly volatile, warming, and resource-constrained environment—namely insufficient fresh water and dwindling soil. The modern food system was built during some of the most stable climate that scientists can observe in the historical record. We have just gotten quite lucky. We have had a relatively stable climate for the past one thousand years, abundant resources of plentiful soil, seemingly endless water, and, more recently, cheap energy. We are now facing a future with dwindling soil, water needs increasingly outstripping supply, more expensive energy, and extreme volatility. Much work remains to prepare our systems to produce a consistent supply of food for a growing population in this set of circumstances. Some areas that are critical to prioritize are as follows:

- Investing in climate-resilient infrastructure such as water management systems, flood protection, and drought-resistant crops.
- Developing early warning systems to give farmers advance notice of potential climate-related risks, helping them adapt in real time.
- Expanding disaster relief programs that help farmers recover after extreme weather events and reduce financial risk.
- Improving and expanding covered agriculture systems to mitigate extreme weather.

By focusing on resilience at the infrastructure level, the government can ensure that farmers and food systems are better prepared to withstand environmental shocks.

## Strengthening Local and Regional Food Systems

Global supply chains have proven vulnerable to disruptions, as evidenced by the COVID-19 pandemic and geopolitical conflicts. To enhance food system resilience, the government should

- support local and regional food production and distribution systems, reducing dependence on long-distance supply chains and ensuring that food is accessible in times of crisis
- create policies that incentivize small-scale, diversified farming to ensure that a wide variety of crops are grown and consumed locally. This reduces the risks associated with monoculture farming, which is vulnerable to pests, disease, and market volatility

## Research and Innovation for Climate-Smart Agriculture

The government can drive innovation by funding research and development in climate-smart agriculture practices. These include the following:

- precision agriculture technologies that improve efficiency in water use, fertilization, and pest control while reducing environmental impact
- development of drought-resistant and pest-resistant crop varieties to ensure food security in changing climates
- carbon farming practices that help sequester carbon in the soil, reduce emissions, and build long-term soil fertility

## Resilience Through Financial Support and Insurance Reform

Financial tools are critical for enabling farmers to adapt to risks. The government should

- reform crop insurance programs to reward climate-friendly practices such as no-till farming, crop rotation, and agroforestry, which build long-term resilience
- expand access to financing for farmers to invest in climate adaptation measures, such as water-efficient irrigation systems and renewable energy for farms

By providing financial incentives for adopting resilient farming methods, the government helps farmers take proactive steps toward long-term sustainability.

## Building a Social Safety Net for Farmers and Workers

A resilient food system depends on the well-being of the workers who produce, process, and distribute food. The government's role in providing a social safety net includes

- expanding access to health care, fair wages, and safe working conditions for agricultural workers, ensuring they can continue to work and adapt to changing environments
- supporting training programs that equip farmers and workers with new skills and knowledge, allowing them to adapt to modern agricultural technologies and climate challenges

## Promoting Equitable Access to Resources

Small and medium-sized farmers, as well as marginalized communities, often bear the brunt of climate impacts and market disruptions. The government must ensure that

- all farmers, regardless of size or location, have access to financial and technical resources to adopt climate-resilient practices
- underserved communities receive targeted support to improve access to healthy, local food, ensuring that adaptation efforts leave no one behind

## Resilience in Food Policy and Trade Agreements

The government also plays a key role in shaping trade policies and international agreements that influence food security. Trade agreements should

- protect domestic food producers by promoting fair competition and ensuring that international imports don't undermine local food production
- incorporate climate and sustainability goals into trade policies to ensure that food systems worldwide are resilient to environmental and economic shocks

This is by no means a comprehensive list of areas where policy can make a big impact. There are countless more that would make a real difference and are worth fighting for. And while we can and must continue to push these policy priorities ahead, in the end I don't think anything I outlined would change how much people like Big Macs, big steaks, sugary drinks, and large portions of all of it. Along with moving the culture and achieving major policy wins, we have to change the companies that are feeding us. The good news is, we can do that.

# Changing Business

Despite the sweeping changes brought about by Healthy, Hunger-Free Kids and Michelle's other programs, the truth is that the government really does have limited influence on how and what Americans eat. Government, as important as it is, is not going to be the cure-all. I get why it's appealing to think there is a legislative solution to fixing all the problems of our current food system: As messy and hard as politics is, so many of our major legislative battles are portrayed in the media (sometimes rightly!) as boiling down to one recalcitrant senator or one heroic congressperson casting the deciding vote that will make history. Sure, that happens, but the appeal of that thought lies in the fact that we want to think that there *is* a silver bullet somewhere—if only we make enough phone calls to that one elected official's office! But the reality is that, for this issue in particular, that's just not going to be the case.

In theory, we could have tried to pass a Healthy, Hunger-Free Act that would apply to every person in the country, not just schoolchildren. But think about it. Should government dictate exactly what crops farmers grow? The animals they raise? How about telling restaurants what they can and cannot serve, or putting strict limitations on salt on your chips or sugar in your cake? Can Congress dictate that home cooks include fruits, vegetables, and whole grains in every meal? There is no way anything like that will ever happen. Nor would it be desirable. Governments change. Do you want someone like Donald Trump insisting that we all eat like him?

And here's the other way to think about it: The food system and the climate are way too big to rely on, or to hope for, the right electoral results to get something done. So what I'm about to say may sound dispiriting, but it's also a way of taking heart when the people you don't want in office end up in office.

For better or worse, it's businesses, the free market, and culture that drive the vast majority of Americans' food choices. Period. If meaningful improvements are the goal, we have to change the huge corporations that exert influence (but really it's stronger than influence; it's basically control) over what and how we eat. The upside is that by working on and with businesses both big and small, we can make improvements that will affect huge swaths of the population in the limited time we have to stave off the worst of climate-induced catastrophe and improve our health.

The downside is that dealing with corporations is messy, fraught, and complicated. No one in our administration relished the idea of working with profiteering executives who most often did not give a damn about the health implications or the environmental ramifications of the products they foisted on the nation. Doing so would put whoever dealt with their likes permanently on the precipice of a PR calamity.

It was going to be a dirty job, but somebody had to try, and I guessed that meant me.

## MARCHING ORDERS

After our successful launch of the White House Garden, the decision was made by the First Lady and our collective team to move forward with a big health campaign for kids and families. Michelle summoned me to her East Wing office and issued some preliminary marching orders. "I care about better food for kids and families, and I want results. Real results. I am here for meaningful and measurable impact for the people that need it the most. You need to make this happen. Figure it out."

I was like, "Okay. Check."

Leaving the meeting, I felt intense pressure and excitement. Michelle was my boss and so much more. I love her as a human. She's a dear friend and I view her as a big sister. There are few people who I am closer to. She was going to bet her reputation on improving the country's food system. And I, a twenty-eight-year-old personal chef, somehow had to help her make that happen. I can't describe how daunting the assignment felt.

No one in any presidential administration had taken these issues on. There was no road map. No playbook. I had no pre-set strategy. To add to the pressure, I was working under the aegis of the nation's first Black president and First Lady, at a time of extraordinary partisanship. This put our every move under a political and media microscope. We had no margin for error. (Recall the outrage around Michelle wearing shorts on a hike in the Grand Canyon, or the "scandal" of the president's tan suit.)

First, we needed to conceive and develop a coherent strategy, and do it on the fly.

As step one, I needed to separate the possible from the impossible. Among achievable goals, we always had to determine which

stood the best chances of success and would make the most difference. Where were our leverage points? Where did change actually
happen? In my case at the time, I had to learn where the government
actually has the ability to either dictate or influence what ultimately
ends up on the nation's plates. And that ability is certainly more limited than I had imagined. It has far less control over food than it does
over other aspects of life.

Take one theoretical example. Conceivably, we could have tried to
enact regulations that set limits and dramatically reduced the amount
of sugar in food products. This would have a big impact on the nation's
health. A lot of our advocate allies would have applauded us. But in
reality, the blowback would have been *massive*. Businesses both big
and small would have been ferocious in their response. This doesn't
just include Hostess and their desire to keep Twinkies super sweet,
but also the local pastry-shop owner making cakes and cookies. Yes,
we could spend all kinds of energy and resources into crafting different carve-outs for different products or companies, but there would
still be an insurmountable challenge to this kind of heavy-handiness:
The American people would have *revolted* at the idea that the federal
government was taking sugar out of their Coke or making the cookies
they love taste worse. At best, our efforts would bog the administration down for years (time we didn't have), and even if we somehow
managed to enact regulations or pass such laws, they would have
been ensnarled in interminable court battles.

We looked at many issues. Ultimately, we concluded that there
were few measures government could realistically institute that would
fundamentally "fix" the food system. To achieve the tangible improvements Michelle wanted, not only did we need to work on strong
policies, but we needed to deal with business—either force it or cajole
it—to make the sorts of changes government could not. Our goal was

to maximize our accomplishments in the limited time we had. Four, or even eight, years is really not a ton of time. I would have to deal with corporations whether I liked it or not. I knew full well that many of my food-advocate friends would despise the idea of it. But when we say we need to change the food system, what we are saying is we need to change the companies that make up the system. When the question shifts from "What should the ideal food system look like?" to "How can we move the system in a better direction and show results?" there is no way around directly engaging the companies themselves. Given the power and role of business, I concluded that failure to deal with corporate titans would have been irresponsible.

Michelle threw down a gauntlet during a 2010 speech she made at a conference of the Grocery Manufacturers Association, the trade organization representing food retailers and major packaged food companies. I had helped write her talk with her speechwriter Sarah Hurwitz. Articulate and poised as ever, she gave an address that was strong and direct and that sent a collective shudder up the spines of those in attendance.

After allowing that government couldn't cure all of the retail food industry's ills by "passing a bunch of laws," she launched into the core of her message:

> We need you not just to tweak around the edges, but to entirely rethink the products that you are offering, the information you provide about these products, and how you market those products to our children.
>
> That starts with revamping or ramping up your efforts to reformulate your products, particularly those aimed at kids, so that they have less fat, salt, sugar, and more of the nutrients that our kids need. . . .
>
> But what it doesn't mean is taking out one problematic

ingredient, only to replace it with another. While decreasing fat is certainly a good thing, replacing it with sugar and salt isn't. And it doesn't mean compensation for high amounts of problematic ingredients with small amounts of beneficial ones—for example, adding a little bit of Vitamin C to a product with lots of sugar, or a gram of fiber to a product with tons of fat doesn't suddenly make those products good for kids.

This isn't about creative ways to market products as healthy. As you know, it's about producing products that actually are healthy. . . .

I know many of you have voluntarily committed to limit your marketing to children, which is a step in the right direction, an important step. And I hope that those of you who haven't will think about doing so as well.

But we have to be honest. . . . We have to ask ourselves, are we really making sufficient progress here? Are we doing everything we can to secure the health and future of our kids?

No one had ever talked to those executives so bluntly. As she spoke, the audience began shifting in their chairs and leaning over and whispering to each other. There was the requisite polite applause at the end, but her words had terrified the attendees. The most influential mom in the country had just told them that they needed to drastically change many of the products they sold, and do it soon. I took a call from Scott Faber, the GMA's top lobbyist at the time, in my office later that afternoon. "When you're going to come at our throats like that, you have to let me know in advance so I'm not caught completely off guard," he pleaded.

Faber (who eventually saw the light and took a position at the Environmental Working Group!) had a point. As a relative newbie to Washington, I had failed to follow the typical administrator/lobbyist way of engagement. After all, while we were going to fight on many issues, I would need to work with them on others if we were to make progress. But despite my faux pas, I think he and members of his organization heard Michelle's message loud and clear.

Michelle and I decided to use a carrot-and-stick approach in our dealings with business. The carrot would be Michelle's highly effective public praise; the stick was the dangling threat of negative publicity and potential regulatory action. I began to explore the corporate landscape. True, there were some companies that really didn't care a whit about the health damage that their business practices were causing and would never budge, and trying to deal with them would have been a colossal waste of precious time. They would continue to sell as much as they could of whatever was most profitable, be it Twinkies or organic carrots. But other companies were disposed to react in a more measured way to Michelle's requests. I had to find out who they were, what they would resist doing, and where they might have room for movement. I also had to consider where in the food system we had opportunities to make the biggest difference.

## HEALTHMART?

On the American grocery retail landscape, there is one elephant: Walmart. I suspect most consumers view Walmart as an assemblage of huge discount stores that also happens to sell groceries. A more accurate corporate profile would be a giant grocery chain that also sells some home goods, clothing, and general merchandise. In the United States, well over half of the company's revenue comes from its food aisles—more than $200 billion a year. It sells more than *one-fourth* of the groceries nationwide, more than the next five largest supermarket chains combined. In many regions, Walmart pockets three out of every four dollars spent on groceries.

But its reach and influence are even broader than these numbers suggest. Here's an example. If you pushed a cart through the aisles of a supermarket in the first decade of this century, you might have noticed a change on the liquid laundry detergent shelves. Suddenly the enormous, multigallon jugs disappeared and were replaced by smaller bottles with labels explaining that their contents were concentrated, and that the more modest packages cleaned as much clothing as their bulky forebears. You might also have noticed that this change swept through every supermarket in the country at once.

The cause was Walmart. Its buyers requested manufacturers to stop diluting their liquid detergent with the usual amount of water, which had been added merely to give shoppers the impression that they were getting more for their money. Walmart publicly trumpeted the environmental benefits of the new containers. The company saved more than 400 million gallons of water, 95 million pounds of plastic, and 125 million pounds of cardboard, which was well and good. But it remained mum about less altruistic motives: the tremendous amount of money the compact bottles saved it—the millions of dollars in transportation, packaging, and storage costs.

Because it made no financial sense for manufacturers to produce smaller detergent bottles for Walmart and to continue making the old larger ones for every other retailer, they concentrated all of their liquid detergent.

There are other famous examples of this. After effective pressure from advocates, in 2015 McDonald's, the largest buyer of eggs in the United States, announced that it would be eliminating from its menu eggs laid by chickens crammed into small cages (as virtually all industrial hens were). Other fast-food chains quickly followed suit. As a result, the percentage of eggs laid by non-caged hens in the United States went from less than 5 percent to *nearly 70 percent*. Reacting to pressure from consumers, more than two hundred restaurant groups and food retailers began phasing out the sale of pork from farms where sows were kept confined to gestation crates, encouraging Smithfield, by far the largest pork producer in the United States, to announce it would eliminate crates from farms it owned. When KFC, Burger King, and other fast-food outlets banned the use of antibiotics important for human medicine in the chicken they bought, Tyson, Perdue, and other major poultry producers stopped administering the drugs to their birds. The corporations' promises were far from perfect (or even entirely accurate), but the point is that the major buyers forced them to recognize the issues and take action industry-wide.

For someone who found himself suddenly tasked with improving the nation's food system, the retail behemoth from Bentonville, Arkansas, truly offered one-stop shopping.

But for a Democratic administration, dealing with Walmart came with a colossal drawback. The company was viewed as the devil incarnate by anyone the least bit concerned with fair labor practices. Famously, a *full-time* Walmart employee could make so little as to still qualify for Medicaid and SNAP. Its practices and reputation have improved slightly since, but even today no one would describe its box stores as bastions of enlightened employee relations.

I had personal reservations about working closely with Walmart. I am a union kid, born and bred. I was too young to remember when I first walked a picket line. My father was an ardent labor activist—a college-educated guy who worked a blue-collar job on an assembly line at Ford, not because he had to or because he enjoyed the work, but because he wanted to advance the interests of the United Auto Workers. When he left Ford to teach, he became president of the union at the University of Chicago's Laboratory Schools.

To honor the United Farm Workers' boycott, Dad forbade grapes to enter our household throughout my childhood. Our fridge never contained chilled cans of Coors beer because of the company's labor policies and the archconservative beliefs of its president, Joseph Coors Sr. We dropped our subscription to the *Chicago Tribune* in the 1980s in solidarity with striking production workers. Some of my earliest memories are of going to rallies, standing on picket lines, and throwing snowballs in the cold Chicago winter at scab workers' cars as they crossed the line.

I would also have to have been an idiot not to know that others within the White House would find the notion of negotiating with Walmart questionable at best, the foremost among them being the president. When he was still a senator running for the Democratic nomination, Barack himself told a forum of the AFL-CIO that he would never shop at Walmart and castigated Hillary Clinton for sitting on the company's board of directors. An Obama campaign staffer labeled Walmart one of the "least labor-union-friendly companies in the country."

But while I was still weighing all these drawbacks, Walmart actually approached me. I was attending a dinner with a group of food company executives in Miami and found myself in a tough debate with bosses of Campbell's soup and Conagra Brands after I firmly pushed them to make their products more healthful.

At a reception following the dinner, I found myself standing around like a wallflower. I was not the most popular attendee, to say the least. A man separated himself from the crowd and, in a thick Scottish burr, introduced himself as Jack Sinclair. He said he was head of grocery sales for Walmart in the United States and informed me that the company was interested in exploring areas where it might possibly work with the administration.

A few days later, I got a call from Leslie Dach, then Walmart's executive vice president of governmental relations and corporate affairs. Dach has an interesting history, one diametrically different from the corporate path of a typical Walmart boss. He was a Democrat through and through, even having worked for the Clinton administration and being active on numerous campaigns over many years.

While at Walmart, his mandate was to burnish the retailer's reputation, which was at an all-time low when he came aboard in 2006. The company's leaders had finally realized that being seen as a bully and prime symbol of corporate malfeasance is not good for business. On the phone, Dach let me know that the company was now aware of issues surrounding its carbon footprint and the healthfulness of the food it sold. "It is one of our top priorities," he said. "We take this seriously. We want to start with health first. Let's talk."

We met for the first time in my office. He was a fit man in his midfifties with curly graying hair and spoke with a careful, halting cadence. He was a thoughtful and direct guy; I could almost see the sentences being carefully formed and edited in his mind before he gave them voice. From his history, I knew that his core values were in the right place, but he was also pragmatic and liked to get things done. During that meeting he said something that stunned me and lent further credence to his expressions of having a sincere desire to work with us. His company's internal studies had shown that fully 70 percent of Walmart's food customers shopped with health as a priority. It was evident that a huge, silent part of the country was connecting the dots between what they ate and their overall health. For Dach, improving the food sold by the company would accomplish two goals. Walmart would be meeting the demands of its shoppers at the same time as it repositioned itself as caring for their well-being. Nothing could drive that point home more forcefully than having the endorsement of Michelle.

I leveled with him. Many in the administration viewed Walmart as untouchable. Unions and their members—some of our core supporters—regarded Walmart as an antilabor monstrosity. It had a reputation for gutting small towns' Main Streets of mom-and-pop businesses—their lifeblood. I told him in no uncertain terms that when word got out that we were working with Walmart, the criticism would be off the charts. And, by the way, I'd be putting my job on the line. We were going to get our asses kicked, no doubt about it. For us to even contemplate taking the inevitable drubbing, Walmart would have to come up with something impactful enough to make the political pain worthwhile. Suggesting cosmetic fixes would be a sure way to end our discussions. Dach was enough of a Washington warhorse to understand those obstacles, but he said that Walmart was indeed prepared to take big steps.

It would take nearly a year and countless messages, emails, phone calls, and face-to-face meetings for us to put together the terms of an agreement that he could take to his superiors and I could bring up to Michelle (who did not involve herself in the minutiae of negotiating). Dach and I discussed issues like reducing sugar and salt levels in the foods sold on Walmart's shelves and bringing the cost of "better for you" products like produce and whole-grain foods into parity with the cost of less healthy white bread, sugary breakfast cereal, and full-fat milk. We explored ways for it to reduce the price of all fruits and vegetables throughout its stores so cash-strapped customers would no longer have to choose between food that was healthy for their families and food they could afford. We eventually agreed on an overall framework that left many details to be dealt with. I did not want to present Michelle with a deal that the top brass at Walmart ultimately turned down, so Dach volunteered to run what we had past his people first.

I was buried in my nightly preparations as personal chef to the First Family, assembling ingredients for the evening's dinner from the kitchen pantry, when Dach called me. He had taken our proposals up to the very top of the corporate ladder. Walmart was aboard.

It was hard to resist doing a fist pump right there in the pantry, which in any case would have been premature. There were still a few intimidating hurdles to get over. I had personally accepted (or at least compartmentalized) the compromises I, the union kid, would have to swallow. I kept reminding myself that my job was to fight food's fight, even if I felt conflicted on labor's fight.

Maintaining the administration's good relationship with labor was the bailiwick of Nate Tamarin, one of the president's top aides. His mandate was to make sure that labor's voice was heard loud and clear during any policy discussions. Nate and I had great professional and personal rapport, but I had been given my own mandate to fix the food system. When I brought up the possibility of a Walmart deal, his eyes widened and he sucked in a deep breath before laying it on the line. We were an administration that had come to power in no small part because of organized labor's support. We needed to do right by them, not applaud companies with horrific worker relations records. Doing any deal with that particular devil would be a horrendous mistake.

I agreed with his points 100 percent—except the mistake part. I tried to explain that the changes Walmart promised to make would help the very working people he was talking about serve their kids healthier meals. Lower prices for more nutritious food would make their lives better, and weren't we an administration dedicated to that goal? I argued that we were not there to sit on our political capital. We were there to spend it on things that were worth the price. And this was worth it. My arguments went nowhere with him; he was doing his job and I respected him for it. We both presented our positions at a meeting with the interim chief of staff, Pete Rouse, and his deputy. They listened and then adjourned.

Shortly afterward, I received a notice that the deal was on. After consideration, Rouse had agreed with me that the good we could accomplish by partnering with Walmart more than compensated for the blowback we would receive from labor leaders. But I'm sure there was another strong undercurrent pushing for the positive decision. He, like everyone else in the White House, was well aware of how important these issues were to the First Lady, so I'm quite sure that there was an element of "Anybody wanna tell Michelle that we've shot down this initiative?" behind the decision to okay the Walmart deal.

There was still one final hurdle to clear: making the arrangement public. If you haven't been in Washington, you can't fully understand what a colossal pain in the ass these "formalities" can be. We had to accommodate about fifty members of the media, along with hundreds of representatives from health organizations, staff from the White House and Walmart, and, most important, Bill Simon, president of Walmart U.S., and Michelle. We wanted to get out our message in a big way (and so did Walmart). It had to be a resounding splash that would allow us to dominate the news with our side of the story before the inevitable naysayers swept in with their messages. We had to send a clear message to other corporations that if they followed Walmart's good example, they too would get to bask in the First Lady's limelight.

At the last minute, events beyond our control forced us to postpone the official rollout. I spent a week on tenterhooks, fearing that someone would leak news of the deal and weaken the force of our announcement. Fortunately, I caught a break. The unveiling went off perfectly.

At the press conference, Walmart announced four key efforts to dramatically improve its food retail program:

- It would reduce salt by 25 percent in packaged foods and cut added sugar by 10 percent. Industrially produced trans fats would be eliminated. In addition to applying immediately to Walmart's house brands, the policy would be rolled out to other national suppliers and food brands sold by the company.

- Customers would save approximately $1 billion a year on lower-cost fruits and vegetables, and Walmart would cut the price premium that often made healthful choices more expensive than less nutritious foods.
- Walmart promised to create a front-of-package seal to help consumers easily identify healthier food options such as whole-grain products and unsweetened canned fruit.
- It would build stores in underserved "food deserts" to provide residents with affordable groceries.

To my mind, the most important change was reductions to sugar and salt because that policy's influence would inevitably spread beyond Walmart and permeate the entire grocery business. As in the laundry detergent example, it would be financially and logistically impossible for manufacturers to make healthier versions of their food products for Walmart and different, sweeter or saltier versions for other retailers. It would mean that basically every grocer in the country would be selling the reformulated products.

Government wouldn't have had a chance in hell of passing a law mandating similar reductions, given the resistance we would have faced from the industry and business-friendly legislators, who as a matter of principle battle every new regulation. Our country, the culture, and our politics simply would not have allowed that to happen. Walmart did not have to worry about any such political consequences. Among its suppliers, the giant's word was law. When the country's most powerful customer strongly encourages a change, suppliers have no choice but to conform, even if reluctantly.

When the arrangement was announced, we did receive criticism from labor activists (including my dad and sister, who phoned me to express their disbelief and disappointment). On the other hand, advocates for better food such as Michael Jacobson of the Center for Science in the Public Interest, who had been stingy with his praise of some of Michelle's other efforts, fulsomely applauded the Walmart initiatives.

I didn't have much time that day to savor our success. It was getting late. As usual, I had to hurry back to the White House to execute my daily quick change from suit coat to chef's jacket and get dinner on the table for the Obamas. I was bent over a cutting board in the small kitchen that served the family residence when from behind me I heard the familiar baritone of the president. He said, "My man!" as he came through the door of the kitchen. "This is exactly why we came to Washington. What you accomplished today will truly help American families. I couldn't be prouder of you," he said as he gave me a Chicago-style handshake and hug.

Barack rarely complimented me on my work and even more rarely offered full-throated praise. He was a great boss, but his standards were at the highest level at all times. At minimum, he *expected* the best from all of us. To this day, thinking about what he said in that moment makes me emotional. I felt insanely good. Here I was, back in my role as a personal chef, hurrying to get a decent dinner on the table for four people, and the president of the United States was saying thanks for what I had accomplished on behalf of millions of Americans.

Buddy Carter, one of the White House's notoriously discreet butlers, came into the kitchen during that encounter. After Obama left the kitchen, Buddy lifted his eyelids in a manner that was in no way discreet. "Da-a-a-a-yam, boy!" he said. "Look at *you*!"

Please understand that I still have plenty of problems with Walmart—and big companies in general. But I also feel that the nation's health and the climate crisis are so critically important that we have no choice but to engage them to change. This is where the game is being played. We should applaud their efforts when they do better in the areas we care about. For all of the abhorrent practices in the industry, our fate rests on our ability to get the companies that feed us to change. This is true both on health and on climate. They have tremendous power to do good, or at least do a better job. We have to deal with that reality. And the news from Bentonville has gotten more encouraging in the intervening years, extending Walmart's influence from better food to combating climate change. Realizing that doing good could also be profitable, Walmart began working with environmental groups in 2016 to explore ways it could reduce its impact. In 2017, Kathleen McLaughlin, the company's vice president and chief sustainability officer, rolled out Project Gigaton, a program to reduce Walmart's greenhouse gas emissions by one billion tons by 2030. More than forty-five hundred of its suppliers climbed aboard. By 2022, McLaughlin reported that the company was already more than halfway toward meeting that goal. The fact is that few people are having a bigger impact on climate issues than Kathleen.

This does not mean that a company doing something positive absolves it of responsibility for practices that are reprehensible. Nor does it mean that once a company makes progress, it will stay committed. Indeed, we are seeing many businesses pull back on, or outright abandon, commitments they previously made about reducing carbon emissions. We must continue to call out companies that are not taking action or are moving in the wrong direction and hold them accountable, just as we should give support to those companies that take risks and try to do the right thing, even if it is flawed.

It's one thing for a popular First Lady to push for corporations to make changes, but everyday people also have a role—perhaps a bigger one than most realize. Certainly, you can vote for politicians who are strong on enacting regulations that will encourage businesses to make climate-friendly changes. You can, and should, call out corporate malefactors on your social media accounts. But the most important thing you can do is to spend your money at establishments that are doing the right thing. This not only encourages them to continue but, perhaps more important, encourages their competitors to take similar steps. Rest assured, when companies see competitors gaining market share by reducing carbon emissions or promoting regenerative agriculture, they will begin making the same efforts. Today this isn't happening at scale. Younger and wealthier consumers are supporting mission-focused brands that are doing better on health and climate. But consumers overall are simply not choosing products that are more climate-friendly. Those who are need to double down on their support and normalize it outside their own social bubbles.

## HAPPIER MEALS

While I had approval from the highest level for the Walmart deal, I decided to go more or less rogue after I initiated discussions with another bogeyman from the world of Big Food: McDonald's. As a group, members of Michelle's staff and I had some preliminary discussions with representatives of the Golden Arches. They brought a large group of people, with a projector and a fancy presentation. My hopes rose with all the pomp and circumstance that they really wanted to change. Their initial proposal to us was a 10 percent sodium reduction in chicken nuggets. To hear them present it, you'd have thought that would solve the world's health and climate change problems. It was a joke, and I told them so. My colleagues came away convinced that the chain had no serious interest in making truly meaningful changes to its menu. Negotiating would be a waste of precious time.

I understood their point, but I also knew there are few companies that have a bigger impact on people's health—especially lower-income people—than McDonald's. So, I kept talks going on a low-key, unofficial level to see if something came of it. Discussions with my McDonald's counterparts were often off premises—at coffee shops, over dinner, or during phone calls. It seemed worth the risk to see if it was possible to get the sorts of concessions from McDonald's that we'd hammered out with Walmart. I knew that if we were serious about improving the diet of Americans, particularly of lower-income Americans, improving the menu at McDonald's would be a giant stride in the right direction.

There was some reason for hope. Like most fast-food purveyors, the world's biggest chain was eager to burnish its image and respond to a growing demand from parents for healthier—or at least less junky—offerings directed at their children. Many adults were also becoming more health conscious about their own eating habits.

The sheer size of the chain is almost beyond comprehension. One anecdote sticks in my mind as an example. McDonald's had done its market research, and surveys showed that customers would readily accept the introduction of a new item, a blackberry smoothie, to the menu. But there was one impediment: There were nowhere near enough blackberries *grown in the entire world* to supply the needs of McDonald's should it roll out the new smoothies. Product managers had to work for two or three years with plant nurseries and berry farmers just to secure an adequate supply of blackberry seeds, to plant enough blackberry bushes to generate the supply for a single new menu item at your local Mickey D's. That's how much impact McDonald's has on the food system.

Despite that unpromising opening McNuggets gambit, I eventually made the trek out to the firm's head offices, then in Oak Brook, Illinois, the site of the legendary Hamburger University campus, where thousands of franchise managers have been taught the gospel of McDonald's founder, Ray Kroc. I, a former line chef at a Michelin-starred restaurant, felt as if I'd entered some sort of gonzo culinary twilight zone.

We did make progress—albeit halting. McDonald's became amenable to cutting sodium, not only in McNuggets but across the board. It agreed to reduce the number of French fries in kids' Happy Meals and replace them with apple slices as the default option. It would also eliminate the sugary caramel sauce that accompanied Apple Dippers. But the company refused to change the drink that came with Happy Meals. It insisted on offering soda as the default beverage instead of water and milk. I told them that there was no way Michelle would ever consider endorsing anything that would lead kids to consume more soda. It proved to be a deal breaker, and the talks collapsed.

Don Thompson, McDonald's president, was furious that the company did not get its Michelle Obama moment and let me know it at a meeting with Valerie Jarrett, a senior adviser to Barack. In an obvious effort to sabotage my position in the administration, he implied that I had not negotiated in good faith during my informal talks with the company. Internal people I had clashed with started circulating rumors around the East Wing that I could not be trusted and that maybe it was time for me to go. Thankfully, Michelle had my back and put a stop to that narrative.

Ultimately, our failed negotiations did bring about positive results, just not through the White House or in a manner that brought the administration any credit. Our discussions had succeed in convincing the company that making healthful gestures could have major positive PR advantages. The following year, McDonald's instituted many of the changes I had insisted it make, including removing soda as the default option in kids' meals. In partnership with Alliance for a Healthier Generation, an organization founded by the American Heart Association and the Clinton Foundation, the fast-food chain more than halved the number of French fries in Happy Meals, pledged to reduce the sodium in all its foods by 15 percent, and got rid of that sugary caramel dipping sauce.

Dealing with McDonald's was super messy. It still serves food that can wreck the health of anyone who dines there frequently, but the hard reality is that the fast-food chain is not going to go out of business anytime soon. Getting it to take meaningful steps, like abolishing that caramel dipping sauce (aren't apples sweet enough already?), can result in real reductions in sugar consumption for the kids with the highest risk of becoming obese and developing diabetes.

As an aside, demand for sliced apples surged nationally by 13 percent in four years, not only because of increased purchases by McDonald's, but because kids liked them and started asking for them in school cafeterias.

## THE NEXT PEPSI GENERATION

To be totally fair, it *is* hard for food business executives to make changes no matter how well meaning they may be. In the early 2000s, Indra Nooyi, then the boss of PepsiCo, was one of only a couple women leading a large food company at the time, and the only woman of color. Nooyi understood that consumers were becoming more concerned about their health and the marketplace was slowly changing as a result. Although she assumed leadership of a company whose core businesses and profits depended on selling sugary drinks and snacks that are full of salt and fat (it owns the chip maker Frito-Lay), she felt it was important to get ahead of the health-conscious trend, and so she set a goal of more than doubling the share of the firm's revenues that came from selling nutritious drinks and snacks. She hired Derek Yach of the World Health Organization to spearhead PepsiCo's global health and agricultural policy, which included working to introduce farmers in developing countries to sustainable practices. She had fat and sugar reduced in some of its core products, and the company added better-for-you fruits and whole grains to others.

For her efforts, Nooyi nearly got fired.

After an earnings call where she didn't mention "blue can" Pepsi once, the knives came out. There was an article on the front page of *The Wall Street Journal* attacking Nooyi, saying she had lost the path. Although she had her eye on the long-term success of the company, many investors' and Wall Street analysts' interests extended no further than what the next quarter's bottom line would look like, to hell with its value five or ten years down the road, let alone its impacts on health or the environment. Nelson Peltz, an activist fund manager, went so far as proposing that the company be split up. Critics accused Nooyi of concentrating on the new, healthier products at the expense of the highly profitable soft drink business. Coke, Pepsi's perpetual archrival, was gaining market share and enjoying an uptick in its stock price. Nooyi, detractors said, was trying to do too much too soon. I'm also sure that her being not only a woman at the head of a Fortune 500 company but a woman of color at the very least subconsciously powered this criticism. After this, Nooyi called me to say she had to step back from pushing on healthier products for some time and focus on selling the core business. She let me know she was going to have to keep her head down for the time being, and from then on she no longer was anywhere near as vocal on the need for the industry to change.

Companies are beholden to consumers and their investors. If we want CEOs to lead their firms in a better direction, we need to work to create conditions to lower the risk of those changes. More energy and attention need to be focused on finance and the role it plays in shaping the behavior of CEOs and company boards. It is of no use if leaders push the envelope on climate change, only to see the stock price fall or themselves get fired.

Wall Street's focus on quarterly earnings deeply embeds short-term thinking in publicly traded companies. If investors don't begin to steer money toward companies that are investing in better food, in food produced in more sustainable ways, and reducing the footprint of their operations, we are going to make only small, incremental progress. I believe large-scale change will start to happen as investors in food and agricultural companies begin to realize how much risk these companies are holding because of climate change. And none of that risk is priced into their stock prices.

Each year we see increasing disruptions to supply chains. In 2024, for instance, drought caused chocolate prices to rise more than 200 percent year over year. Think of the challenges for a chocolate company's performance if its key ingredient sees that kind of increase in price year over year. Governments and activists need to focus on shaping capital markets to think longer term through policy and public pressure. We must support companies that are investing in making changes to their operations around climate and health.

There are some signs that movement is beginning to happen on the climate side. Many of the big food and agriculture companies have made pledges to cut their emissions dramatically, some putting aggressive targets by the end of the decade, others by midcentury. Many of these companies are piloting programs and some are showing signs of progress. One example is a partnership Unilever made with growers in Arkansas that supply most of the rice for its Knorr brands. Working with the University of Arkansas and Arkansas rice grower Mark Isbell of Isbell Farms, Unilever helped promote growing methods that enabled rice crops to spend far less time flooded, which reduces bacterial growth that causes large methane releases. As a result, methane emissions for this crop dropped by 76 percent, and overall greenhouse gas emissions by 48 percent. This is a significant impact and a model that can be replicated globally. Without a doubt, meaningful progress is possible.

You can be assured that the company's executives, however well intentioned, are not being purely altruistic. The most forward-thinking leaders realize that climate change could seriously disrupt their supply chains and are taking steps now to blunt the worst scenarios. They are also aware of the possibility (albeit distant in the United States) that government regulations may someday force their hands. They value getting ahead of the curve, especially if it can be done with limited financial downside.

While there are signs of progress, most of the early efforts are not worth highlighting and none of it has really begun to scale. Most big companies' efforts are anemic at best thus far. Even the companies that have invested initial resources to try to change their supply chains are being questioned internally on the return on these investments, particularly in the face of consumers not spending dollars in the mass market for more sustainable products. Not dissimilar to governments, in many of the big companies there are small groups of people pushing for change who have to fight against most of the leaders who are solely focused on growing their revenue and increasing their profit margins. Even in the more progressive companies, there are big internal battles about whether to invest real dollars in changing supply chains at scale.

While there is reason to have some hope because of the actions of a handful of leading companies, the actions by the industry so far as a whole have been woefully small both in funding and in scale. Part of the reason is these companies have not been built to be good at this. They have built their businesses on the ability to create food at the lowest price with the highest margin and getting us to want to eat it. The expertise required to start fundamentally overhauling supply chains to reduce environmental impact is entirely outside the expertise and incentive structure of the companies. Investing in this critical transformation generally creates little short-term value for them. It is not helping them sell more product, take market share, increase their sales velocities, or reduce their costs. Subsequently, budgets for most of the companies' sustainability teams are an afterthought, and the teams are small. Companies need to get better at figuring out how to decarbonize. To get better, they will need to invest exponentially more into building teams and expertise and then give those teams the resources and tools necessary to figure out the best solutions. And ultimately, if we want the companies that provide the vast majority of the calories we consume to produce those calories with a more regenerative approach, we need to value that way of production and pay for it at the store; we need to show them that this will help these companies outperform ones that don't.

Good-food activists often tell me that tinkering with sugar, salt, and fat percentages by companies like Walmart, McDonald's, and PepsiCo, or incrementally cutting emissions, will never fix the food system. It's not reform we need, they say, but revolution. Theoretically, I agree with them, in that small changes by food industry behemoths won't get us all the way to where we want to go. Nor do I think positive actions by big players absolve those same players of other business practices that are damaging to society. But to think that we could force them to go away through some sort of foodie revolution is not serious thinking. Billions of people have to be fed every day. The industry has trillions of dollars invested in the status quo. They are not going to pack their bags and leave because we don't like how they are doing things.

I see no other path but to engage these companies, beat the shit out of them when they are not aggressively taking on climate and health, praise them when they make a genuine effort, work to push investors to take into account these other issues when they invest, and help create more and more demand for better products. There is no shortcut on changing this complex, deeply rooted system and the businesses that constitute it.

There is, however, one key sector of food production where revolution *is* the norm: agriculture. I'm convinced that a technological revolution will be the key if we're to help the food system avert catastrophe. Many of the new and better tools and innovations that are being developed will help companies make big changes to their operations and enable businesses to increasingly find solutions on climate issues.

# Changing Technology

Picture this: Farmers can barely grow enough season to season, year to year for themselves and their families, leaving little or nothing to export to hungry city dwellers. Prices skyrocket for the food that *is* available. What were once everyday necessities become luxuries available to the wealthy few. The multitudes who cannot afford enough to eat resort to theft and violence. Or simply starve. Humanity doesn't go extinct but limps along, caught for generations in this cycle of food insecurity, scratching for sustenance, sometimes successfully, sometimes not.

Conditions seem irredeemably bleak, but suddenly a new technology revolutionizes agriculture. Within a short time, crop yields double, then *triple*. Individual farmers, able to cultivate much more land by using the new technology, can grow enough to sell excess crops for cash to buy equipment that makes them even more productive. This creates a whole category of new off-the-farm jobs for those who manufacture and maintain that equipment, as well as those who process and sell the crops. Soon, each grower can feed three to four times as many people as they did before the arrival of the new technology, breaking the cycle of food insecurity. And a better-fed society goes on to enjoy centuries of growth and prosperity.

This scenario is not a fairy tale or a work of science fiction. Nor is it futuristic. I am describing factual events that transpired in Europe a thousand years ago. The miraculous technological breakthrough that in all likelihood saved European civilization: the plow.

Before the plow, farmers cultivated their fields with devices called ards, not much more elaborate than sharpened sticks pulled by draft animals to scratch shallow troughs in the soil in which to bury seeds. As a result, fields were poorly prepared to nurture crops, and the crops that did grow provided meager yields. Farmers tried to amend the soil with manure, but it stayed on the surface, where it was washed away by rain, providing little fertility. Using an ard was slow, inefficient work that did little to inhibit the growth of weeds. Plows, which flip the soil, bury vegetation that would otherwise compete with crops and open the ground to receive manure. It seems simple, maybe even obvious now, but for centuries life before the plow was mostly a battle against starvation—often a battle that many didn't win.

## AGRICULTURE IS REVOLUTION

I don't know exactly how long it took between the first person fashioning a simple plow and it becoming a ubiquitous, world-changing tool. But if your impression is that farmers are a conservative group, not keen on change, well, they have reason to be. On any given day, they manage an enormous amount of risk: heat, frost, drought, pests of many types, bacteria, fungus and a plethora of microorganisms, input costs, market prices, and on and on. Adding any new variable into their operation comes with inherent risk. They also have a tremendous investment in the status quo, in all of the systems and equipment that make their operation go. As long as things are going well or even acceptably, okay. Making changes could invite catastrophe. "If the way Dad and Grandpa did it works, why take the chance of doing it another way?"

That logic takes deep root if you're working on such slight profit margins (as most farmers are) that a misstep could spell crop failure and financial disaster. Agricultural families have passed down more than just practices or preferences for a particular brand of tractor or variety of seed. Like all other sectors of society—only more so—the force of cultural norms in farming is powerful. I have spoken to many farmers who have adopted the latest climate-friendly regenerative methods; many of them spoke with me about the isolation they have felt from their communities and sometimes families for going against the dominant culture of the best ways to manage their operations. I have heard stories of ridicule and growers being ostracized as they transitioned their operations to a regenerative system. A no-till field may be the best thing for the soil and climate, but to some traditionalists it looks "messy" compared with a "clean," fully tilled one.

But there is a flip side. When farmers encounter a new invention or a novel way of doing things that *really* works—that adds more value than risk—they have shown an eagerness to adopt it at tremendous speed, making the history of agriculture a story of periods of stasis interrupted by revolutions.

Ever since the first nomads opted to take the revolutionary step of planting the seeds of wild grains they gathered, the history of agriculture has been a chronicle of a series of revolutions, with one new, interruptive technology after another bursting onto the scene and allowing humans to produce dramatically more food, frequently doubling output, and often just in the nick of time.

The invention of agriculture itself was a revolution that permitted nomadic hunters and gatherers to settle in permanent communities.

Early Middle Eastern civilizations further boosted food production through ingenious irrigation systems that brought water to rainless but fertile river valleys. A little over a hundred years ago, technology once again transformed farming by replacing draft animals with machines. After World War II, newly bred plant varieties, artificial fertilizers, and chemical pesticides led to a geometric growth in production; this in turn alleviated famine in some areas of the developing world.

Technological revolutions are the very essence of agriculture. In many ways they *are* agriculture. They have not only changed the way food is produced but transformed entire societies. Yet we tend to be leery of technological changes in the way food is produced. It's understandable and not without wisdom. We tend to be leery of most changes in what we eat; for good evolutionary reasons, it's wise to stick to the safe and familiar. And not all innovations have been good for society; some incredibly efficient hyper-processed food technologies, for instance, have been downright bad for people. And others have had serious unintended consequences. But clinging to a traditional way of producing food and fearing change just because we're resistant to change itself seem downright silly when you take a historical perspective.

In that light, saying we already know enough to raise food in an environmentally sound manner by using tried-and-true methods of the past such as natural manures and heirloom seeds—practices I employ in my own vegetable garden—is largely irrelevant to feeding the global population in an economically viable way, however beautiful it may be to contemplate. This common opinion robs agriculture of its single most animating feature. Yes, looking to the past for wisdom is valuable. But we are facing new challenges in climate and in scale that our ancestors did not develop answers for. We need to look for new ways to apply principles of the past, and new tools will help us attain new goals. Those new tools and models should draw on historical wisdom but should not be constricted to some idyllic image of the past or calls to go back. It's simply not possible to feed the world in our climate-changed present and future exclusively with techniques that have existed in the past.

We are on the verge of another revolution in food production that could be as profound as the invention of the plow. Given the challenges climate poses, this revolution is coming at a time when agriculture desperately needs another eleventh-hour innovation. We should both recognize and encourage the truly beneficial new tools: new ways of operating farms, new scientific breakthroughs, new ways for farmers to finance their operations. Combined with the best of age-old practices, they will form the fourth pillar upon which we build the food systems of the future: innovation and technology.

After leaving the White House at the end of 2014, I launched a new career. My job today is discovering the entrepreneurs who are pushing these new tools into the mainstream of food production as a partner at Acre Venture Partners, a mission-driven investment fund looking to solve the issues of climate change, environmental health, and human health through food and agriculture systems. Our job is to help young companies gain access to the necessary investment and markets that are essential to make their visions realities—for all of us.

We see an endless number of start-ups trying to tackle a broad range of challenges. In telling you what I am seeing in this space, and the exciting vision some of these innovators have, I hope to show the role innovation can play in creating the tools we need to help solve the challenges before us. I'm about to introduce you to some companies I have invested in or worked with, as well as others that I have no formal relationship with, which I believe hold key pieces of the solution. I realize there's a risk in this book of coming off as boosterish and self-serving. But I'll take that risk. I know many of these organizations intimately, and the people who operate them. I have seen the results of their innovations in real life—not just theory. I am convinced that they—and many other like-minded businesses—could be playing key roles in bringing about the next agricultural revolution, the one we need if we are going to survive the climate crisis. I only work with companies that I truly believe can have a measurable and meaningful impact on the issues we care about. By no means are these companies our only shot at getting this right, but the companies that I highlight are the kinds of companies I believe hold a key piece of transforming our food system. Honestly, by the time you read this, some of their CEOs likely will have been replaced, and other companies will likely have ceased operations; that's just how things go when you're launching something new and trying to tackle hard problems. New companies face an endless array of challenges; they may have young, inexperienced founders, harsh competition from other start-ups, and investors who often are just in it to make money and maximize their return even if it comes at the detriment of the company. I have experienced this firsthand, and it is devastating to watch, knowing how the potential of a company can be extinguished by shortsighted, possibly nefarious investors. Young companies also have to compete with large corporations, which have massive scale and economic advantages; in food and agriculture in particular, the large consolidated incumbents wield a lot of power. But I believe deeply in the missions and the ideas of these upstarts. They are not only creating new tools to solve some of our challenges but also working to shift the culture of business; as they grow in success, they pressure the big players to change, adapt, and improve. They are really a driving force for change on many levels.

Unlike many in the venture funding world, I am not a blind technologist. I do not think all of the innovations that have been created have benefited society. Nor do I believe that technology alone can solve our problems or is the magic bullet to climate change. On the other hand, there are some challenges in our system that will absolutely require new tools to overcome. That's a fact.

Yet we all have observed continual knee-jerk reactions to new technology in the food space. In some cases they amount to moralistic demonizing of new approaches. I am often asked whether I think a certain technology is good or bad, but my response to that is usually, "That is the wrong question." Tools are inherently neither good nor bad. A hammer can be used to either build a home or cause severe bodily harm. The questions we should be asking ourselves are these: How will the tool be used? Who benefits? And what are possible unintended consequences if an innovation achieves scale? If a company is using a new plant genetic breeding tool to improve nutrient density, enhance flavor, reduce the need for synthetic chemicals to grow it, and resist drought, then I am likely a supporter of that tool. If a plant genetic breeding tool is used to apply more chemicals to the environment and siphon off profits from growers to a large corporation, then I am not a supporter. Sometimes the answer is black and white; other times it's more in the gray and comes with trade-offs. We should all try to answer these questions with an honest assessment of how challenging the path ahead will be. And it all has to come back to a real, honest look at the future we are trying to leave for our children and their children.

## THE CARNIVORE'S DILEMMA

Don't get me wrong. I enjoy meat as much as any other red-blooded son of the American Midwest. Few things make me happier than simply coming home from work on a pleasant summer Friday evening, firing up the grill, and slapping on a couple of nicely marbled rib eyes for the family. But I also know what the science says. I know that one of the most important acts each of us can do to reduce our greenhouse gas footprint is also one of the easiest—at least in principle: We must cut down on the amount of animal protein we consume, *particularly* beef.

Some experts estimate that cattle ranching in its modern form alone generates two-thirds of agriculture's contribution to greenhouse gas emissions, thanks to the prodigious quantities of methane the animals belch, the emissions from fertilizers used to grow the feed for the cows, and the vast tracts of once-carbon-sequestering rainforest cleared to create pastureland. It's a double-bad scenario: One of the most effective ecosystems in taking carbon out of the atmosphere is burned down to make pasture for raising billions of animals that literally burp the most potent greenhouse gas.

It's vital that we cut back on how often we indulge in those backyard barbecues or stops at fast-food outlets to grab a couple of quick, inexpensive cheeseburgers. Notice that I wrote "cut back," not "cut out." Let's face it, most Americans—I count myself among them—are just not about to eliminate meat entirely from their tables anytime soon, or soon enough to address climate change. I'm all in favor of eating meat from animals that are grass fed and/or pasture raised according to regenerative practices, both of which can reduce cows' contribution to environmental damage. But that means of production, by nature slower and smaller scale, is more expensive and is not going to get us where we have to go on its own. Currently, grass-finished beef accounts for less than 4 percent of U.S. consumption. It also costs about 50 percent more than conventional feedlot beef, and many consumers find its leaner, firm texture less desirable. I enjoy eating grass-fed beef myself, but the majority of meat eaters simply are not going to alter their flavor preferences or pay more money for the sake of greenhouse gas reduction.

And it's not just a case of rich nations cutting back on their animal protein consumption. Research shows that any reductions achieved in today's developed countries will be offset by increases in worldwide meat consumption as residents of developing countries become more affluent and able to afford the beef, pork, and chicken that Americans eat as if a birthright. Given the reality of increased meat consumption, we are going to have to find other ways to reduce the footprint of livestock while simultaneously working to reduce overall consumption in developed nations. This means reducing the emissions of the feed (more on that later) and reducing the methane released by the animal. There are many companies working on additives that do reduce methane. But no matter what, the planet cannot support a population set to go to ten billion, along with billions of cows, pigs, and chickens.

All of which explains the race to develop meat analogues that are produced without involving a cow, chicken, or pig. The holy grail in this race is a food that (1) costs no more than "real" meat and poultry, (2) delivers the same taste and mouthfeel, and (3) contains similar amounts of protein and other nutrients not readily available from plants but abundant in animal flesh.

We aren't nearly there yet, but two companies blazed the trail for public and commercial acceptance of plant-based meat substitutes. In 2009, Ethan Brown, an environmentalist and businessman, founded Beyond Meat in the Los Angeles area. Two years after that, the Stanford University biochemistry professor Patrick Brown (no relation) launched Impossible Foods in Silicon Valley. Rather than relying on plant products cooked and pressed into patties, they've created proprietary methods for actually creating meat-like protein from plants. It's a pretty amazing technology, really; Impossible in particular has found a way to create the red-tinged protein found in blood and, well, meat, and that gives the product meatier flavor.

Today, these meat substitutes are widely available through supermarkets and fast-food franchises, pushing the category in the United States to $1.4 billion in annual sales. Impossible and Beyond absolutely deserve great credit for being first movers in popularizing alternatives to animal protein. They've smartly marketed through chefs as well as fast-food and other restaurant companies with massive reach, and quickly done what might have seemed, well, impossible: They've begun to shift the culture—remember the first pillar?—and have begun to normalize the idea of eating meat made without animals.

It absolutely helps that to most meat-eating consumers these products taste way better, with texture far more meat-like, than the veggie burgers that preceded them, which remind me of over-salted and over-spiced beans. Those early meat substitutes bore little resemblance to actual beef and were mostly for die-hard vegetarians and vegans pining for something, anything, to put into a bun.

Despite their breakthroughs in taste and texture, Beyond and Impossible have real limitations in the big picture. The most obvious is that the processes they've developed so far produce only ground products: hamburger, sausage, chicken nuggets. If you have a hankering for a grilled sirloin or a seared chicken breast, you're out of luck. And that *really* matters, because ground beef is a by-product, not the product. Think about it: As long as there is demand for steaks, cows will be raised, slaughtered, and processed for the twelve or so rib eyes you get per animal. Ground beef is what you make after you've processed and sold all the cuts that can be sold as more expensive and more valued steaks and roasts. So a plant-based replacement for the *by-product* of steaks won't realistically make a dent in the actual amount of meat we produce.

And at a time when thoughtful eaters are trying to avoid highly processed foods for health reasons in favor of natural ones, an Impossible Burger qualifies as ultra-processed in the extreme, containing twenty-one ingredients, including textured wheat protein, coconut oil, potato protein, natural flavors, leghemoglobin (from soybeans), yeast extract, salt, konjac gum, xanthan gum, soy protein isolate, vitamin E, vitamin C, thiamin, zinc, niacin, vitamin $B_6$, riboflavin, and vitamin $B_{12}$. Beyond Burger goes one better with twenty-two ingredients. Conventional hamburger meat contains one ingredient: beef.

Plus, many alternative protein products use pea protein, which, unadorned, tastes awful (comparisons vary from cardboard to chalk), so the manufacturers have to rely heavily on additives to mask the off-flavors before they can even begin to layer on beefy or chickeny taste. A patty of faux burger packs no less fat than regular ground beef but five times as much sodium. The results cost twice as much per pound as real meat.

## The Meat of the Matter

Despite the introduction of plant-based alternatives, sales of real beef have remained steady with slight fluctuations year over year. Meanwhile, Impossible, Beyond, and other plant-based meats are running into commercial headwinds. Their sales in grocery stores tumbled 14 percent by volume during 2022. Restaurant sales were down 9 percent over the same period. So, while these pioneering companies have made an important cultural contribution by carving out a place for plant-based meat substitutes on the nation's menu—a major step forward—the new products are not likely to take us far on our journey toward curtailing our meat habit. "What remains looks more like a niche category than a meaningful replacement of an entrenched industry," a Bloomberg article in 2023 stated. But in the battle against climate change, the goal should be not to disparage niche categories that have the potential of reducing greenhouse gas emissions but to build upon them until they achieve a critical mass.

My partners and I had very little interest in investing in meat analogues. Early on, we had seen prospectuses from just about every company that came to the market. None of the products checked the fundamental boxes of health, great taste, and real sustainability. To put it another way, I hadn't seen a product I'd serve to my kids. Then I got an email from Lucas Mann, one of my partners: "I think I may have found a good one."

He included a link to the site of a company. I clicked and immediately knew this company was different. The home page was an image of what looked exactly like a steak, with a beautifully seared exterior and a mouthwatering rosy interior. From the photograph, I couldn't detect any difference between it and any sirloin I would proudly pull off my own grill—an immediate distinction from the ground-beef-like products dominating the category.

But any teenage nerd with rudimentary computer skills can enhance an online photograph. So, I checked the product's nutritional profile, and this is when I knew the product was in a different league. The ingredient list I saw was refreshingly brief compared with the twenty-plus ingredients of other meat substitutes. It contained only four items: mycelium (the rootlike fibers of mushrooms and other fungi), acacia gum (a natural product made from the seeds of a plant grown in India and Pakistan, commonly used to give food products texture), salt, and less than 2 percent natural flavorings.

The nutritional profile showed that a slice of this "meat" was virtually a superfood, combining all the dietary benefits of meat and plants. A four-ounce piece delivers seventeen grams of protein, 34 percent of the daily recommended value for adults, and, unlike most plant-based meat products, it provided a complete protein package, supplying all the essential amino acids. That same serving provided *29 percent* of the daily recommended amount of dietary fiber. If you know me at all, you know that I am a fiber nut; in fact, when I look at product nutrition labels, fiber is really the only positive attribute I look for. Fiber benefits the body—and just as important, the microbiome in our guts—probably more than anything else on the nutrition label. It's why every doctor and nutritionist is always telling you to eat more whole grains. The vast majority of Americans do not consume anywhere near the recommended daily allowance of fiber (less than half, in fact). Beef, pork, chicken, and fish provide no dietary fiber at all. And here was a "steak" that gives you nearly a third of the RDA.

The product was also rich in riboflavin, niacin, and vitamins $B_5$, $B_9$, and $B_{12}$. What it lacked was equally impressive: no saturated fats, no cholesterol, no starch, no sugar. Basically, this alternative protein combined the beneficial micronutrients and fiber of plants with the amino acids of meat. I thought, "Holy shit! I have never seen a food like this."

But all these qualities would have meant nothing if the stuff tasted like rubber. I'd never to my recollection eaten mycelium. But I knew enough about the restaurant and food-service business to understand that the general public would turn their noses up at a product, however good for them and the planet, that wasn't also delicious.

The company was called Meati, based in Boulder, Colorado. This was at the peak of the pandemic, and like everyone I had hunkered down in full cocoon mode. But the chef in me was not going to invest a penny in a meat alternative that didn't have the mouthfeel and taste of a cut of the real thing. So, I sucked it up, strapped on a mask, and headed out to the airport for my first flight in the COVID era. JFK airport was abandoned—no one standing in front of the check-in counters, no lines at the TSA inspection station, no waiting as I grabbed a coffee from a lonely-looking Starbucks barista. In Denver, I picked up a rental car and drove west toward the mountains to meet a couple of my partners, who also had flown in to have a taste for themselves.

## Where's the Mycelium?

Now, before I get you too excited about Meati, you should know that in the time between when I first wrote this and the final edits, the company has since filed for bankruptcy, leaving its future as a commercial operation in serious jeopardy. It's likely to be an incredibly disappointing outcome—caused, I believe, by some mistakes by management, compounded by some poor and ethically questionable investor practices. And it serves as an object lesson in how hard it is to bring real global solutions to scale.

But what I saw in my initial research into Meati is a superfood technology they helped pioneer that I believe holds real, potentially game-changing promise. And even after Meati's downfall as a business, there are already multiple new companies (I've counted at least eight) working toward bringing this product to market. I have little doubt that mycelium will be an important source of nutrition in the future, so I still want to share with you what I saw in Meati—and my enthusiasm for this area.

After I got off my flight to see them in Colorado, my GPS led me to the outskirts of Boulder, to a dreary industrial park of warehouse-like buildings with nothing but numbers to distinguish one from the other.

Tyler Huggins, Meati's co-founder, made up for the mountain-town ambience his headquarters lacked. About five years out of a PhD program at the University of Colorado (and still looking every inch a grad student), he greeted me like a bro, smiling and pumping my hand. His cheeks showed signs of not having felt a razor in a couple of days. His brown hair was slightly disheveled, swept behind his ears and allowed to flop over his open collar. He'd rolled up the sleeves of a light blue cotton shirt, and his feet bore a pair of hiking sandals. His vibe suggested that he'd just hurried back from scrambling down a mountain trail or intended to go out on one as soon as he'd finished dealing with the moneyman from New York. Until then, he would be warm, at times humorous, and exude zero pretense.

As our first order of business, we headed out to a patio where a charcoal grill was already alight under the supervision of Justin Whiteley, the company's soft-spoken and thoughtful co-founder. He was tall and had a trim dark beard. Whiteley worked quietly behind the scenes overseeing scientific research and the nitty-gritty of building the capacity to produce products in quantities sufficient to supply a national market.

Huggins slipped one of their faux-chicken breasts (the company featured them as well as beef-like "steaks") on the coals to let it sizzle. And sizzle it did, sending the unmistakable aromas of live-fire grilling into the air while he tossed a basic green salad, adding slices of the cutlet once it was done. That was it. No added flavor steps or flourishes. We sat together outside sipping a local brew on a beautiful, fiercely blue Boulder afternoon with the peaks of the iconic reddish-brown Flatirons as a backdrop. In that setting, we tucked into a meal that could revolutionize the way humans eat.

Whiteley had yet to fine-tune the production process, but the "chicken" in my salad showed tremendous promise. Even in that beta stage of development, it put all the existing commercial meat substitutes to shame. The slices were nice and juicy and had a quiet umami, that savory taste that other alternative proteins often lack. It had a mouthfeel and chewiness that wasn't perfect, but came really pretty close to a chicken cutlet, and delivered a pleasantly neutral flavor profile, which was the goal. The product was like a blank canvas. Cooks could take the cutlet in any direction: southern fried, barbecue, jerk, Cajun, you name it. Diced, it would go great in a bowl of noodle soup or a potpie. I felt that Whiteley needed to make some tweaks (maybe reduce that chewiness a bit), but the cutlet had few other obvious flaws. It was instantly clear that their offering was different and had vast potential.

At first, a trip into Meati's production facilities seemed somewhat underwhelming: Huggins led me into a space jammed with floor-to-ceiling stainless-steel tanks, valves, gauges, metal ladders, and a tangle of hoses and pipes. It didn't look, shall we say, like the cutting-edge start of something that is supposed to change the way the world eats. The place could have belonged to any college town brewery, except the youthful crew that ran it wore white lab coats and hairnets instead of jeans and plaid shirts. Whiteley explained that my first impression had merit. They used jerry-rigged brewery equipment to create its paradigm-shattering products, because they're fermented rather than engineered. And that is actually a significant part of the promise: We won't need cutting-edge tech to make this stuff.

The starting point (and main ingredient, at 95 percent) is mycelium. The naturally filamentous structure of mycelium makes it well suited to a variety of applications. In addition to food production, it lends itself to the manufacture of faux leather, cloth, and (because its fibers conduct electricity well) batteries. As grad students at the University of Colorado, Huggins and Whiteley set out to determine whether mycelium's conductivity would make it a suitable component for an electric battery that would maintain its capacity to be recharged far more times than lithium-ion and other batteries currently in production. They succeeded, sort of. Their experimental mycelium batteries performed as hoped, but failed to attract the interest of manufacturers.

The silver lining for Huggins and Whiteley was that in developing the battery, they became familiar with many species of mycelium, some of which were edible, highly nutritious, and had no off-putting taste. That fibrous structure effectively mimicked the texture and mouthfeel of muscle fibers in meat—a trait existing alternative meat producers who rely on plant-based proteins have found almost impossible to replicate. Mycelium could easily become the base not for ground "meat" but for realistic steaks, chops, and cutlets: the White Whale of all meat alternatives. Huggins and Whiteley also felt that a truly effective and delicious meat substitute would have a bigger environmental impact than another battery technology.

The mycelium strain is placed into the tanks along with sugar and water to nourish it. The resulting fermentation causes the fibers to grow rapidly; in fact, this particular strain of mycelium is one of the fastest-growing organisms in the world. According to Whiteley, the strain they use doubles its volume every two hours. Left alone, a morsel the size of a sugar cube can grow to be the size of a chicken breast overnight. Once scaled up with the right equipment and staffing, Meati planned to be able to yield an amount of "meat" *equivalent to forty-five hundred cows every day,* only without the high labor costs and waste of a slaughterhouse (no hides, bones, inedible viscera, or feathers). And, of course, with minimal impact on soil and water, and almost no emissions of climate-altering methane and carbon. Huggins claimed the process is "over a thousand times more efficient per acre as far as resource use than conventional meat production. And even compared to something like soy, we're well over twenty times more efficient at producing protein per acre using significantly less land, water, and energy in the process."

Seeing the game-changing ramifications of this new product entering the marketplace, other investors joined us and fell all over one another wanting to put money into the company, eventually fueling its growth with nearly $300 million in funding. We helped bring in a wide array of chefs and experienced executives to help get the company ready for their launch. Questions do remain about whether the public would embrace a product whose main ingredient is fungal fibers, even though humans have consumed them throughout history (mushrooms are the reproductive parts—the "fruit"—of mycelium, much as apples are of apple trees), and the Food and Drug Administration has already determined that such products are completely safe. In early 2022, Meati gave its chicken cutlets the ultimate test by offering them for sale to its email subscribers, who bought more than a thousand in the first hour. The entire inventory was gone in a day. The company made deals to sell both chicken and steak substitutes through Sprouts Farmers Market (a nationwide chain of natural food supermarkets), Whole Foods Market, the restaurant chain Sweetgreen, and others.

Huggins's goal was to make products that sell at the same prices as conventional beef and chicken. He wants shoppers to view them as sources of everyday protein. "Ultimately, we are introducing a whole new type of protein. We're realists. I don't think we will replace meat, but it could supplement it and be a way of offsetting increasing demand for animal proteins by giving customers an alternative."

He's right. This food won't be a full meat replacement even if it becomes a massive success. But what is clear is that mycelium should be a mainstay of our diets in the future. It checks all the boxes. And the more new foods we develop with similar characteristics, the more we will be able to take a sizable bite out of meat production's contribution to climate change, give us more possibilities to feed ourselves in an increasingly difficult world to produce food, and improve people's health along the way. Meati's own future may be uncertain, but I am rooting hard (no pun intended) for all its competitors now.

## Building a Better Cow

It is clear we need to reduce the amount of meat consumption through replacements and a shift in how much meat we like to consume. While we need to continue to work toward that goal, the reality is we are going to continue to consume an immense amount of animal protein for the foreseeable future, including beef and dairy. That raises the question: What if we could find a way to have a more environmentally friendly cow? Today's beef cattle are agriculture's number one producer of greenhouse gas, responsible for something like 18 percent (estimates vary all over the place) of global emissions. Blame a simple fact of biology: Cattle (and sheep) ferment grasses and other fodder in their guts to make them digestible. Methane is a natural byproduct of bovine fermentation, and cows burp it into the air.

Advocates of grass-fed and regenerative cattle-raising practices are sometimes fond of proclaiming, "It's not the cow, but the how." They lay the blame for cattle's contribution to climate change on massive cattle operations that rely on finishing cows not in pastures but in huge feedlots and on a diet heavy in corn and other grains—the infamous CAFOs. There is doubtless a good deal of truth to this position. They also point out that manure builds soil health, and with the right grazing practices cattle raising can build soil health and sequester some carbon. It is unclear if raising cows on well-managed pastures actually produces lower emissions than traditional feedlots. This is due to the fact that the lifespan of a feedlot cow is shorter, because it puts on weight faster due to the grain diet—resulting in fewer of those burps over a lifetime. In either system, recent research has also uncovered feed additives derived from seaweed, fungi, or other sources that could reduce the emissions impact of beef and dairy production. But catchy rhymes aside, ultimately you can't get around that key biological reality. Cattle, no matter how they are raised, produce too much methane. So, sorry, it *is* the cow.

Which explains why an Australian start-up called Nbryo has focused on developing a breeding program for herds of cattle and cows with a much lighter environmental impact—including methane emissions—than current ones. According to Nick Cameron, Nbryo's co-founder, the company is using bio-digital and robotic technologies to supercharge husbandry practices that have existed since the 1970s, but until now were used only on a modest scale.

Not all cows are equal offenders when it comes to gaseous emissions (like humans in that regard, I suppose). According to Cameron, a case study of one New Zealand dairy herd found that individual lifetime methane emissions varied from more than thirty tons of carbon emissions to less than ten tons per cow. Using traditional breeding methods, a farmer could in theory gradually reduce the methane emissions of the herd by breeding the cows that emitted the least methane and culling the others. But that's a painfully slow process—a cow produces but one calf per year.

Cows get pregnant in a few ways. There is the good, old-fashioned way: Put a bull out in the pasture with the cows and let them do their thing. There is artificial insemination, where semen is injected into the cow, or there is embryo transfer—essentially IVF. The traditional way leaves genetics to chance. Artificial insemination—the other AI—is a bit more expensive but offers the ability to get some better genetics and more diversity into the herd. But generally, the genetic gains are slow year over year. In vitro fertilization allows for much more control of the genetic qualities of the offspring, because breeders have full control over which eggs and semen to combine and ultimately which embryos to implant. With our ability to do genetic screening, we can naturally select for genes for cows that are more heat tolerant, for instance, or that emit less methane. The challenge is that right now this method is very expensive and is used only in elite programs to breed bulls or Wagyu beef, which famously costs upward of hundreds of dollars per pound and is purely a luxury food. But what if this technology can be made affordable enough that a typical producer could use it and see genetic gains for their herd year over year?

Nbryo's suite of technologies dramatically shortens and improves upon traditional methods. It is working on the whole process to try, at every turn, to reduce the cost of creating and implanting a healthy embryo. This starts with how traits are identified, all the way to developing a device to easily implant the embryo, which is currently a difficult process requiring a highly trained technician. It is working with producers and genetic companies to take their top genetics and bring those to the entire herd in one generation. Embryos can be selected for optimal feed conversion, reduced water requirements, and heat tolerance, desirable traits in a world of climate change. Sexual selection of embryos can also eliminate the problem of male dairy calves, which are typically culled and mostly sent for dog food, with a small amount supplying the tiny veal market, but only after their mothers have spent a year carrying them, eating, drinking, and burping methane the whole time.

Working in cooperation with five universities in the United States and Australia, and with investment from the Bill and Melinda Gates Foundation and an Australian meat-marketing organization, Nbryo has conducted extensive field trials. It expects to improve and simplify the process so that any farmer can breed a herd of bovines that consumes 17 percent less feed and emits 26 percent less methane than today's.

Such methods enable Nbryo to make improvements in a herd in seven days that would normally take a farmer seven years. Farmers will be able to select the most desirable animals at the embryonic stage, unconstrained by the genetics of their current herd. There are no Frankenstein insertions of foreign genes here; it's simply taking a breeding technology that exists (for human and animal alike) and improving on it to enable its benefits to be affordable for a much larger portion of the market. The biggest potential for impact is in developing nations like India, where the animals are far less productive than in other parts of the world. For families subsiding on a cow or two, this improvement can make a material difference in their economic security and their quality of life.

## GREEN GENES

The problem of meat production and climate is a massive one, but the production of plant foods is an arena where we need major advances, too.

Ponsi Trivisvavet's career-changing epiphany came while she visited an impoverished village on an Indonesian island to attend a harvest festival. Syngenta, the seed and agrochemical multinational where she was an executive at the time, had recently developed a corn hybrid suitable to growing conditions in Southeast Asia, and she had gone to the village in part to see firsthand how it was doing. A peasant farmer, whom Trivisvavet guessed to be in his eighties, approached her and began to thank her profusely. By growing the new seeds, he claimed to have nearly doubled the output on his small plot. For the first time in his life, he said, he was able to sell enough surplus corn to afford to buy beef for his family. "Once something like that hits you, it stays with you," Trivisvavet told me. She pledged to dedicate her career to creating a better food system.

When I first met Trivisvavet at the Cambridge, Massachusetts, offices of Inari Agriculture, the seed development company she now heads, I had a vague understanding of the new gene-editing technology called CRISPR (pronounced "crisper"), which exploded onto the scene in 2012 after a landmark paper was published demonstrating that the technology could edit the genetic material of a plant. Some of what I knew was downright scary, most notably reports about a Chinese researcher who used CRISPR to engineer human embryos born with modified genes that resisted the virus that causes AIDS. That alone would be a good thing. But unrestrained by legal and ethical guidelines, CRISPR could lead to parents (or governments) being able to order up designer children at will. That's frightening. A CRISPR-like technology had also gotten some iffy publicity when it led to the development of potatoes and apples that resisted browning after being cut. Such cosmetic changes to crops are all well and good, but hardly momentous advances in view of the challenges we face in trying to feed the world of the future.

But after a couple of hours in Trivisvavet's office, CRISPR's potential appeared much more exciting—and practical. I came away convinced that CRISPR will profoundly change the food system. I'll go so far as to say that thirty years from now virtually all the major food crops grown will be shaped by CRISPR.

To label Trivisvavet as a powerhouse is an understatement. Born in Thailand and raised in California, she worked for more than a decade at Syngenta, a giant $28 billion seed and agricultural chemical company that operates in ninety countries around the world. Trivisvavet rose steadily to the position of president of the company's American seed operations, but in the years following her aha moment with the Indonesian farmer she came to feel that Syngenta had become moribund and wedded to traditional GMO products.

As CEO of Inari (named after the Shinto god of rice, fertility, and success), she set out to combine the power of CRISPR and that of artificial intelligence with the ultimate goal of lightening the footprint of agriculture on the environment while providing better profits for farmers.

In many ways Trivisvavet defies all the stereotypes of a tech start-up entrepreneur, and her company has little in common with those Silicon Valley firms with their roots in the garages of their founders' parents. She is in her late forties and dresses in conservative, well-tailored pantsuits. I have rarely met anyone with her no-nonsense focus who is also a genuinely kind and good person.

Under her leadership, Inari has attracted more than $750 million in investment capital since its 2016 launch. With facilities in Cambridge, Massachusetts; Ghent, Belgium; and West Lafayette, Indiana (the home of Purdue University's respected agriculture department), the company has grown to more than 260 employees.

The other aspect that truly differentiates Trivisvavet is that she has assembled a team with credentials that rival and even surpass her own. Catherine Feuillet, Inari's chief science officer, is a veteran of Bayer Crop Science and the French National Institute of Agricultural Research and is in her own league when it comes to plant genetics. Inari's scientific advisory board literally comprises the leading researchers on this technology including Jennifer Doudna, who won the 2020 Nobel Prize in Chemistry for her revolutionary work on CRISPR. A common trait of this team is a desire to take what they had learned about this industry and use it to have a real impact, in a way they never could at big, incumbent agriculture companies. Many of these brilliant minds had become disillusioned where they were and set out on a new path to make a difference.

## A Primer on CRISPR

In simple terms, CRISPR technology enables scientists to more quickly, inexpensively, and accurately alter or "edit" DNA. (I'll get into how it works, which is key, in a bit.) Today, the technology is being used in the development of huge commodity crops such as corn and soybeans, to make them far more resistant to drought and able to produce higher yields than current varieties, on far less land, and with smaller amounts of fertilizer and water. These are the same crops that have for years been modified by GMO techniques.

There are several key differences between current GMO techniques and CRISPR (or, I should say, "yesterday's GMOs," because CRISPR is making GMOs as we know them obsolete). One isn't about the technology itself: It's who gets to wield it, and for what purpose. Because of its efficiency and lower cost, the technology of CRISPR lends itself to use by companies with more modest budgets than the massive ag companies, including start-ups that could not previously compete in plant genetics. The technology can also be used to improve less commercially valuable non-commodity crops such as fruits and vegetables. These areas have often been overlooked because they were smaller and the huge expense of innovation meant that investing in these areas could not be justified. Ultimately, these smaller companies in smaller markets have the potential to play an important role in providing highly nutritious foods with low environmental impact for marginalized humans who desperately need access to healthy fare. Only time will tell if that becomes true, but there is real reason to hope these tools will help us provide more sustainable, nutrient-dense foods in the challenging years ahead.

## CRISPR Versus GMO

A common misconception is to lump CRISPR together with standard GMO techniques. They do share one major trait: Both processes alter the genetic expression of a plant or animal. But that's where the similarity ends. Traditional GMO work typically involves introducing genetic material from one species into another, entirely different organism—say, inserting a bacterium into corn that produces a chemical poisonous to caterpillars, something that would often not happen in nature.

In most cases, CRISPR activates or deactivates genes *already naturally native* to an organism's genetics, as opposed to inserting genes from foreign species. Given enough time, money, and good fortune, traditional plant breeders could, in theory, encourage the same traits. In essence, CRISPR hyper-accelerates what farmers have done for thousands of years when they save seeds from the best plants—the most delicious, say, or the most productive, or the most disease-resistant—to be sown the next season. It's just that, instead of having to select those desired traits, find the plants that express them, breed them, and hope for the best, a researcher using CRISPR can find the places in the genome where, say, a strawberry gets its sweetness. In a variety that has tremendous yields, those genes might be inactive, resulting in tons of berries that taste like, well, nothing. But flip that genetic switch, and watch the next generation bear huge yields, with fruit delivering the full flavor wattage of a handpicked summer berry. Because the genes involved are in strawberries anyway, whether expressed or not, the U.S. Food and Drug Administration has decided that it is not necessary to regulate the introduction of CRISPR crops. The assumption is that because they contain no foreign DNA and could have been developed through traditional breeding techniques, the stringent regulations governing traditional GMOs need not apply.

But even if you have no qualms about the safety of traditional GMO technology, I would argue that if you are an eater, there can be no doubt that it has been a massive disappointment. Although virtually all corn, soybeans, and some other commodity crops growing today are genetically modified, the technology has never come close to living up to its initial billing. After billions of dollars of investment and decades of research and development, the vast majority of GMO crops do one of two things: Either they produce chemicals that kill certain insect pests, or they can survive applications of herbicides, allowing farmers to spray those chemicals to kill weeds while leaving the GMO plants unharmed.

Developing GMO crops is a slow process, taking a decade or more to create a single variety with one new trait and then an equal amount of time to get that variety approved by the FDA. Given the pace of climate change, that is time we simply do not have. The cost and complexity of cumbersome GMO techniques means that only huge companies such as Monsanto can afford to invest in the technology. It also means that utilization of this technology made financial sense on only a few commodity crops like corn and soybeans because of the size of those markets. This has resulted in tremendous efficiency to the underpinnings of the hyper-processed food industry, and on a relative basis made whole foods like fruits and vegetables increasingly more expensive.

One Inari scientist sees the difference between traditional GMO techniques and CRISPR as akin to the difference between using a sharpened rock and using a surgical scalpel to perform complicated surgery. With CRISPR, technicians can select precisely which genes they want to alter and aim directly at them, as opposed to blindly injecting DNA and hoping some lands where it will deliver the desired

trait. Critically, CRISPR can be used to make multiple edits simultaneously, and researchers see the results after a single generation of
offspring. CRISPR is also faster and relatively inexpensive. Using it,
scientists can develop new crop varieties in a couple of years at one-
tenth the cost of traditional GMO work.

But advocates will still have to introduce CRISPR to the public in
the proper way, which means doing the opposite of what big companies did with GMOs back in the mid-1990s. Back then, the seed giants
basically rolled out the modified crops on the sly, keeping the public
in the dark. Their condescending attitude was "You don't need to know
about this stuff. Trust us." They succeeded only in building skepticism. Agribusiness compounded this error by paying researchers—
also on the sly—to write articles and make media appearances
extolling the virtues of bioengineered crops and scornfully dismiss-
ing possible downsides. This practice backfired spectacularly when
media reports brought it to light.

To further demolish any goodwill, teams of corporate lawyers sued
struggling farmers for violations of GMO patents—like the famously
outrageous case where some unfortunate farmer was sued by a mega-
corporation after the company's seeds got blown onto his field. Or
when they hauled food manufacturers into court who dared to label
their products as GMO-free, even if the statement was true. Scientists
and journalists who expressed concerns saw their reputations ruth-
lessly attacked by agribusinesses and their allies. To many consumers,
bioengineering became viewed as pure evil. If CRISPR is to be
accepted, it must be rolled out honestly and transparently. And this
has to be done before the widespread introduction of CRISPR crops.

Ultimately, consumers have never really benefited in any direct
way from the GMO revolution. All the benefits have gone to the companies that hold the patents on GMO seeds. They profit mightily from
seed and chemical sales. Large growers also benefit—at least in the

short term. By planting GMO crops, they can control weeds and insects with less effort and cost. With virtually nothing in it for us, why should the public be interested in shouldering any GMO risks?

## A Better Tomato?

Take tomatoes as an example. Commercial seed companies have relentlessly focused on breeding tomato plants that produce increasingly large harvests of tough-skinned fruit that can survive cross-country truck trips and retain plenty of their eye appeal in supermarket produce aisles. By those measures, the modern industrial tomato has been a runaway success. A tomato plant today is three times more productive than one grown in the 1970s. A tomato pulled from that plant can tumble off a semi and maintain its perfect near-spherical shape. Alas, along the way, traits that produce important heart-healthy and cancer-preventing nutrients such as vitamin C and lycopene—not to mention traits that deliver decent, let alone great, flavor—have been lost. Why bother with the time and expense of breeding for health and taste when the grocery retailers and fast-food joints that are your primary customers won't pay you a single penny more for a half-decent tomato?

CRISPR, however, enables breeders to reintroduce the genes that express those beneficial traits into bigger, tougher, commercial tomatoes. Although much of the development remains in laboratories and greenhouses belonging to researchers and has yet to actually reach farms and gardens, the new technology has been used thus far to develop varieties that resist devastating infections. Tomatoes hate cold weather; CRISPRed versions can tolerate cooler temperatures, which will be important as we face more extreme weather, which will include early frosts. Some have a long shelf life, without sacrificing color, taste, and nutrition. Thanks to CRISPR, farmers can grow stout new dwarf varieties that stand up to high winds better than gangly conventional plants.

Other examples abound. Citruses with resistance to cankers and bananas that have immunity to streak virus (a great boon to farmers in impoverished tropical regions, who depend on the crop) are both products of CRISPR. If you enjoy chocolate, be thankful that CRISPRed cacao plants can survive the lethal ring spot mosaic virus.

CRISPR researchers have developed rice varieties whose yields are 25 to 30 percent higher than today's, crucial for a crop that feeds more than half the world. Better strawberries, grapes, apples, cucumbers, lettuce, mushrooms, and potatoes have all resulted from CRISPR, although they are currently still in the research phase. And CRISPR is being used to develop a variety of corn that, in addition to providing higher yields, locks carbon from the atmosphere safely in the soil. Given the massive quantity of corn grown around the world, if we could use corn to sequester carbon, that could represent an important step toward reducing atmospheric greenhouse gases.

## CRISPR on Steroids

Let's hope CRISPR is introduced to the public properly. Because it can do so much good and it's getting even better. At Inari, Trivisvavet's team has supercharged the technology via a two-step process the company calls SEEDesign. The first step involves deploying machine learning, or AI, to develop a deep understanding of the gene flow in a plant. This supplants time-consuming methods in conventional breeding, which basically required growing hundreds of plants and having horticulturists examine each to see which carried a sought-after trait, then breeding and crossbreeding those plants for generations until a line was developed that consistently produced the desired trait. Breeders were basically wandering blindfolded through a maze. "The technology allows us to understand the interconnectivity between genes and the pathways genes use to interact with one another," Trivisvavet says. The result is an electronic "blueprint." Off come the blindfolds to reveal a clearly marked path. Then comes the second step: Researchers use the blueprints to guide the CRISPR-based tools directly to the genes that express the most useful characteristics of a crop. Inari then sells the resulting "parent" seeds to seed companies that breed them out for sale to individual farmers. "Our mission is to design seeds that produce plants that benefit farmers, consumers, and the environment by requiring less land, fertilizer, and water," said Trivisvavet.

Using only genes that are native to the plant species they work with, Inari currently focuses on three crops: corn, soybeans, and wheat. While going for non-bruising apples or tasty tomatoes would have some consumer benefits, Trivisvavet chose Inari's first crops because of the potential environmental impact. Combined, they occupy nearly 70 percent of the earth's total cultivated land. This focus allows the company to maximize the impact of its crops. "CRISPR is a way to unlock the real potential of the genes of a plant. That might happen with traditional plant breeding, but it would take a thousand lifetimes," she said.

The crops that Inari has in development promise to have 10 to 20 percent higher yield than current varieties. Under current plant-breeding techniques, yield increases are limping along at 1 percent. Inari's CRISPRed crops will deliver these yields using 40 percent less fertilizer, which is responsible for about 2 percent of humans' greenhouse gas emissions, about the same amount as the entire country of Germany. And they will absorb 40 percent less water, which is critical if we are to grow crops under hotter, drier climate conditions. Importantly, the new crops are already growing and on farms, according to Trivisvavet, and will be ready for general usage within two years. And good thing, too, because her estimates suggest that the time to boost production is limited. "This has to happen well before the end of this decade," said Trivisvavet.

TRUE, A LOT OF THE talk around CRISPR sounds similar to the promise of GMOs, and I fully understand why many would remain skeptical. That skepticism is healthy. The jury is still out on how this technology will evolve, and ultimately who benefits. But given the tsunami of challenges that are coming with climate change, it is unwise and even irresponsible to oppose CRISPR based on principle alone, as some anti-GMO groups already do. We will need every effective tool at our disposal to ensure that everyone has nutritious food produced in a way that does not destroy the planet. I think CRISPR can play a major role. If the technology is effective, we should embrace it and be vigilant about how it is deployed and ultimately to whose benefit.

## THE SOIL WILL SAVE US (WITH A LITTLE HELP)

Innovations such as those made by Inari could help humanity take giant steps toward reducing the environmental impact of some of our most important crops. But to get where we need to go—in the time frame the science says we have—we are going to need technologies that dramatically increase our ability to sequester carbon on a large scale. The most promising solution I have seen is a combination of two processes that has been successfully taking carbon out of the atmosphere for the last 400 million years or so: the natural actions of plants and fungi.

There is plenty of evidence that we are beginning to reduce

greenhouse gas emissions, and it's beginning to make a difference. But no one could credibly argue we are doing enough to actually meet the moment. Societal change worldwide is happening, but it is simply happening too slowly. We are not reducing enough emissions to protect the ecological health of the planet in time. The UN Framework Convention on Climate Change stated clearly that job number one is for countries to reduce emissions at the source. But in 2023, the UN repeated that only 15 percent of the reductions agreed upon in the 2015 Paris accords were on track to meet goals set for 2030.

As recently as the 1970s, reductions in emissions alone would have been enough to prevent the dire effects of global warming. But we have been too late in coming to the game. Reducing the carbon we spew into the air will *no longer get us where we need to go on its own*. Even if we miraculously slash carbon emissions in half in the near future (something no sensible person would bet on), humans have already pumped far too much $CO_2$ into the atmosphere. Today's levels top 400 parts per million (ppm), the highest it's been in 400,000 years, an era well before the first small tribes of *Homo sapiens* evolved in Africa.

In the late eighteenth century, just prior to the Industrial Revolution, the $CO_2$ level hovered around 280 ppm. We built the entirety of modern human civilization—culture, technology, and especially the agricultural systems that feed us—to function effectively between 280 and 350 ppm. Yet, historically, that range is something of a rare $CO_2$ sweet spot for the earth's atmosphere, neither too much carbon nor too little for the current life-forms on planet Earth. We breached its upper limits four decades ago and blithely continue along on our polluting ways.

Turning the climate clock back to that Goldilocks era requires not only that we stop emitting carbon into the air but that we actually remove meaningful amounts of it that humanity has already put there.

It's a challenging task, but doable—at least in theory. Unfortunately, theory is not good enough, given our current predicament. It is hard to overstate the urgency of the moment—where we are already 75 percent of the way to an increase of two degrees Celsius, when disasters are going to get really wild.

Many ideas, examples, and even prototypes abound, some more or less promising than others. Researchers and entrepreneurs have tinkered with several so-called negative emissions technologies that take carbon out of the atmosphere and put it somewhere where it does not contribute to climate change. Two companies, Carbon Engineering in British Columbia and Climeworks in Switzerland, use a process called direct air capture. As the name indicates, the companies deploy giant fanlike machines to suck air into chambers where it mixes with chemicals that bind with the $CO_2$. Those chemicals are then treated in a way that causes them to release their carbon, which is in turn captured.

A second process known as bioenergy with carbon capture and storage is a fancified version of raising plants and trees, which draw carbon out of the air as they grow, and then using them as fuel in power plants. Scrubbers much like those in the smokestacks of industrial polluters today would then remove and sequester the carbon from these exhaust gases.

Both approaches sound great, but the technology is absolutely not ready for prime time.

Some suggest that the warming effects of climate change can be countered with so-called geoengineering, which involves having aircraft seed the upper atmosphere with sulfur dioxide to create clouds to block some of the sun's rays, thereby cooling the earth, not unlike the effects of massive volcanic eruptions. Others propose sending what are basically massive umbrellas into space to cast a shadow back onto the earth to cool it down. But geoengineering faces staggering obstacles. Mainly, these are almost all at best Band-Aid solutions that do nothing to reduce the $CO_2$ already in the atmosphere. Cloud seeding would cost billions initially and would have to be repeated frequently as the sulfur settles back to earth, creating acid rain as a nasty by-product. And geoengineering's effects would be difficult to control. What might be good for one region of the globe could be disastrous for another.

And all of these technologies have the same set of problems. Currently, they all would be prohibitively expensive to scale up. Tens of billions of research dollars have already been spent with little to show for the massive investment. Their effectiveness is questionable. And they remain too small to make a dent in the climate problem. Climeworks, for example, charges in the ballpark of $1,300 per ton of carbon.

One researcher compared these approaches to trying to bail out a sinking boat with a thimble. I am not arguing that these kinds of innovations are not worth pursuing; it's possible that at some point in the future one of these methods could in fact be vital to humanity. The main point is that none of them will be operating at scale soon enough to sequester adequate carbon in a time frame that meets the urgency of the moment according to the science. I haven't heard of any technology for removing $CO_2$ from the atmosphere that has a chance of being ready for widespread introduction within the next five years, if then.

So that leaves us with green plants and their partner microbes. Through the natural process of photosynthesis, plants inhale $CO_2$ and combine its carbon with water to create the sugars they use for energy. And they do a damn good job, absorbing about a quarter of humanity's current carbon emissions. About half of that $CO_2$ is respired almost immediately back into the air, but some is sequestered by the plant itself to produce roots, stems, branches, leaves, and fruits.

Carbon not needed immediately by the plant gets pumped down into the surrounding soil, where it is fed on by microbes. These microbes in turn make available to the plant essential nutrients it needs for its growth.

Because of modern farming practices like excessive plowing, leaving fields bare of vegetation in the offseason, and applying chemical fertilizers, today's agricultural soils have released 60 percent of their original carbon into the atmosphere, according to the UN's Intergovernmental Panel on Climate Change, the definitive global research body on climate. A hundred and ten million metric tons of carbon in our atmosphere were once in our agricultural soils. That is eighty years of humanity's current carbon footprint. It's a big part of how agriculture has contributed to climate change. But the fact that soil has such an enormous capacity to store carbon, and that plants and microbes can sequester carbon in the soil, is one of humanity's great hopes for minimizing the impacts of climate change.

## REGENERATIVE AGRICULTURE, AND WHY IT'S SO IMPORTANT

One possible route to getting carbon back into the soil is through widespread adoption of regenerative farming. A regenerative approach to agriculture goes far beyond the organic methods that were once considered the gold standard of conscientious farming. As the name implies, regenerative agricultural systems aim to regenerate soil and its surrounding ecology to its pre-chemical biodiversity and health, including taking carbon out of the air and fixing it back in the earth. Such an approach has multiple benefits, including taking carbon dioxide out of the atmosphere and burying it in the ground, where it does not contribute to climate change. Regenerative farming does so by harnessing complex and powerful natural processes of microbial life in the soil and the ecology of the natural systems around it.

That said, regenerative farming is not a specific protocol. No one has a clear definition of "regenerative farming," and the phrase means different things to different people. To some it is a way of life. To others it is a set of values, a philosophical approach to producing food that seeks to build, not just extract. It focuses on the holistic health of the soil and the ecosystem that is the foundation of a resilient food system. At its fullest expression, it focuses on the well-being of not only the soil but the animals and people that are a part of that ecosystem. It is rooted in not only seeing value in what we extract but also recognizing the inherent value in the ecosystems that produce our foods and goods.

To me, though, a lot of time and energy is being wasted on debating what is and what isn't regenerative: What practices constitute it? How to standardize it? How to label it? And so on. I understand the urge to do so, but ultimately, if we are not careful, this can become a real distraction. The focus should be solely on outcomes. Are we lowering the footprint from agriculture? Are we sequestering carbon in the soil? Are we reducing water use? Are we building soil health? Are we increasing beneficial plant, insect, and animal populations? These are the outcomes that will ultimately determine our ability to feed ourselves. Frankly, I am open to just about any tool, practice, or technique that accomplishes these outcomes.

The benefits of the regenerative approach are well documented. Large companies are now beginning to make investments to transition their supply chains. Farmers who have transitioned from chemical-reliant mono-cropping report a long list of benefits, including higher profitability, lower input costs, increased water efficiency, improved soil health including increased carbon storage, and so on. I will describe an operation that fits this bill later in the book, but the question we have to ask ourselves is, If this is a better way to produce food, why isn't everyone farming this way? Or more accurately, Why is only a tiny amount of land farmed this way?

Currently, about 1.5 percent of arable land in the United States is being farmed regeneratively. If our future and our kids' future depend on making this transition, what is needed to bring it to scale?

The harsh reality is that this general approach is not creating enough value in the marketplace for growers to take on the inherent risk and cost of change. Simply put, if the operational changes on the farm were risk-free and there was near-term financial benefit, farmers would do it en masse. Yes, there are visionary producers who have figured out a way to make the leap and forge a different path. There are others like Gabe Brown, whom I will profile in the next chapter, who was forced into this way of growing because he faced insolvency, couldn't buy conventional inputs, and ended up seeing the benefit of a new way of working. But while these examples show us what is possible, they give us little insight into how we get broad-scale adoption of a different way of growing.

For most farmers, the risk-reward calculation doesn't pencil. Transitioning an operation means using different techniques, different practices, and potentially different crops. All of these changes come with significant risk. Remember, growers are managing more risk on a daily basis than most of us deal with in a year. And given that farms operate on very low margins, there is no room for self-inflicted mistakes that could cause calamity. There is also real cost to making these changes. Practices like cover cropping mean a whole new season of seed and passes on the tractor to plant that seed. Everything has a cost. This leads to the other side of the challenge. It generally takes three to four years before growers start to see any yield gains from improved soil health. This means they have to take on all the risk and cost, with the incremental benefits coming years later. This reality does not add up for most growers.

So what could the answer be to shift the calculation? There are a number of different levers that could be pulled to make the decision to transition operations to a more regenerative way, and I support deploying all of them. They include better costs of financing capital, better insurance products, and better policy incentives (though, as we know, this last point is probably going to be lagging). We will need a multitude of answers. But the core question should be, How can growers create the most ecological benefit at the lowest cost, and who would be willing to pay for what? There is a real cost to this transition, and if we want growers to change, we have to figure out who is going to cover the cost.

Remember when I said there are lots of benefits to regenerative farming? Improved water efficiency, better biodiversity, that whole long list? None of those should be taken lightly, and they all hold tremendous value to human existence. The market has to start putting a price on that value. (More on that later.) But the most urgent societal need, and the area the market is starting to value, is the promise of reducing carbon in the atmosphere. That's what drives almost all investment in this area. But there are issues.

Some scientists are skeptical that plants can fix enough carbon in the soil naturally to move the needle under current farming methods. And there are serious questions about how permanently that carbon stays in the soil. Accurately measuring that carbon is not easy to do cheaply. We're not able to create enough value for growers to adopt these changes. Yet.

# AUSSIES TO THE RESCUE

Recently, a group of Australian entrepreneurs and scientists developed the most exciting answer to the carbon conundrum that I have come across. It's been literally right under our feet all along. The company they formed in 2018 (and which my partners and I backed financially), Loam Bio, has developed a means to dramatically increase carbon sequestration in common crops, including corn and soybeans, by exploiting the eons-old symbiotic partnership between plants and soil-dwelling fungi. In addition to sequestering carbon, Loam Bio's technique increases soil fertility, boosts plant growth, prevents disease, protects against drought and hot temperature, and improves soil's ability to retain water, making it the very definition of a win-win proposition.

Initially working in greenhouses, the Loam team identified and analyzed fifteen hundred species of a naturally occurring fungus belonging to a group called dark septate endophytes. These microbes, which attach themselves to the roots of crops, help plants absorb vital nutrients such as nitrogen. In exchange, the fungi receive carbon-rich sugars from the host plants, which they use as nourishment and eventually exude into the soil as pure carbon. Normally, most of the carbon stored in the earth as organic matter by plants reenters the atmosphere through decay, cultivation, and other practices that disturb soil. The length of time it remains in the ground varies widely, depending on soil characteristics, drainage, climate, and farm practices. But septate endophytes convert sugars into two kinds of stable carbon molecules: One is called mineral-associated carbon, which is permanently stored in the soil, and the second type is stored in minuscule balls of soil called microaggregates. In either form, the carbon is said to be fixed. It will remain trapped underground for hundreds if not thousands of years, according to Guy Hudson, Loam Bio's co-founder and CEO.

Hudson is deeply compassionate and has spent a career committed to fighting climate change at a number of nonprofits and NGOs. With a bushy red beard framing a deceptively boyish face, he is blessed with the upbeat, can-do attitude so often associated with his fellow Australians. He combines these traits with a seemingly infinite appetite for straight talk and hard work. His co-founder Tegan Nock has a warm, salt-of-the-earth spirit and is a third-generation farmer. Her grounded demeanor could lead some to underestimate her depth of intelligence. Tegs, as she is affectionately called, shapes the company from the lab to the fields with a deep understanding of the way producers operate and how the agricultural system works. If the technology doesn't meet growers where they are, she knows adoption will be slow and the impact minimal.

In contrast to so many entrepreneurs, Hudson and Nock are extremely collaborative, enabling them to manage an ethnically diverse group consisting of about three dozen farmers, biologists, and financial experts at four laboratories in Australia and the United States. The company delivers its product to growers by coating seeds in the beneficial fungi. Seed coating is an inexpensive practice commonly used in other agricultural applications. Farmers planting Loam Bio's treated seeds can employ the same equipment and practices that they already do.

By 2022, Loam's treated seeds had left experimental greenhouses and been planted in pilot trials, covering four thousand acres. Early results are nothing short of stunning. "Our studies show a 7 to 17 percent soil carbon increase over a season," said Hudson. "If you extrapolate that out to the 1.8 billion hectares that humans have in crops each year, you would be looking at eight gigatons of $CO_2$ a year." That equates to about one-quarter of $CO_2$ emitted by humans. "The entire annual U.S. aviation industry's emissions could be removed if our seed coating was applied to America's soybean crop," he said. For farmers, it could represent a financial windfall as well.

Part of the potential power of Loam's technology is that it is system agnostic. Loam's method works in conventional farming systems that till and use fungicides, herbicides, and the like. For a grower, the risk to use this tool is very low to none. They don't have to shift anything they are doing to start investing in carbon and better soil health. Additionally, the payback can come that first year, from yield increases and carbon credits, and the cost of the product is less than what a grower would make on the increased yield alone. This technology changes the risk and return on investment calculation. Carbon supercharges plants' growth and improves their overall health. Loam's trials have consistently shown significant yield increase, and in addition the soil has many secondary benefits like increased drought tolerance— win, win, win. Once producers begin to see financial benefits from ecosystem services like carbon sequestration, they will be strongly incentivized to begin adopting more regenerative practices. And they will move to optimize their operations' ability to sequester more and more carbon.

But back to the fundamental question: Who is going to pay for it?

## WHAT ARE CARBON CREDITS, AND DO WE NEED THEM?

The most obvious calls are for big food companies to pay significantly more for products that are produced in an ecologically sound way. Some progressive companies will pay incrementally more (some already are), but they simply have not and will not pay enough to incentivize widespread adoption. Alternatively, others say the government should pay. Both have a major responsibility. But given the state of politics in the United States, there is little governmental appetite for investment that comes close to what would be needed to pay for this transition and sustain it.

But if farmers will be helping to solve the greatest threat humanity has faced, shouldn't they be paid well as a result? The only path that enables the valuing of the service that producers can provide is to allow them to sell carbon credits, both as insets and as offsets.

Selling, or trading, carbon credits has origins that go back to the 1990s, when nations came together and introduced so-called cap-and-trade policies in an attempt to reduce the sulfur emissions that caused acid rain. In the United States, government placed a limit, or cap, on the amount of sulfur dioxide ($SO_2$) individual power plants could emit. Those whose emissions exceeded the limit got taxed on the excess. Conversely, those that produced less than their limits could sell the remainder of their allowances to those that exceeded their allotments (allowing them to avoid taxes). This system created a financial reward for plants with environmentally friendly practices— a carrot-and-stick approach. These market-based incentives worked. $SO_2$ emissions had fallen by more than 40 percent by the mid-'00s and have continued to fall.

The two hundred nations at the annual UN climate summit in 2021 agreed to form a similar market for carbon dioxide. Attendees created carbon credits, which permitted companies or other organizations to emit a specific amount of $CO_2$ (one ton of $CO_2$ per credit). Today, markets for carbon have become well established. Credits can be bought and sold on exchanges in more than forty countries. Importantly, the $CO_2$ markets encourage polluters to purchase carbon offsets, based on credits earned by fishers, forestry companies, and farmers for the carbon they sequester. By buying these credits, airlines, trucking firms, factories, and technology companies can offset their overall carbon footprint. Farmers and others who store more carbon than they emit can make a profit from selling the credits.

With the election of President Joe Biden in 2020, the USDA dramatically accelerated its efforts to bolster agricultural climate solutions. Robert Bonnie, Biden's undersecretary for farm production and conservation at the agency, has a long history of encouraging ecologically sound practices on farms, both from inside the government and from his decade-long career with the Environmental Defense Fund, a conservation group.

One of his first moves in his new position was to propose the USDA's Partnerships for Carbon-Smart Commodities program. This program spent just over $3 billion to help pay for partnerships throughout the system to support cover cropping, proper rotation of grazing livestock, no-till practices, and other activities. These resources have spurred a lot of activity on the ground, gotten many companies engaged on the transition, and were a powerful step in the right direction. Think of it as a very large pilot project made up of more than 140 programs across the country, involving production of both animals and crops. This is the first step to figuring out how to make change make financial sense to growers.

Image-conscious private corporations have already created demand for carbon credits to offset their carbon footprints. The pressures of possible government regulations and public opinion (and in some cases, the realization that climate change could be bad for profits) mean the demand is only going to grow. Many companies have set decarbonization targets and are looking to show progress toward those goals. If we can maintain momentum, demand should continue to grow (albeit more slowly at the moment absent the Trump administration's support), and increased pricing for carbon should follow, further incentivizing growers to get in the game. In early 2021, carbon was worth between $2 and $15 per ton. By early 2024, high quality credits were trading for around $80 a ton. The Canadian government has set up a carbon marketplace that is targeting prices approaching $100 a ton. Greater financial rewards will encourage more producers to consider "farmed" carbon an important addition to their operations' revenue.

While there was significant hype and excitement in 2021 and 2022 for carbon markets, that enthusiasm has waned—in part because of intense criticism for how they practically function from many I admire in the environmental movement. Articles in *The Guardian,* for instance, attacked many of the forestry-based credits—the majority of the current market—as complete bunk: Companies are able to buy credits, which are ostensibly used to protect existing forests. But many of these forests are not under threat of being cut down. And even for forests that are under threat, those credits are considered "avoided emissions," meaning the credit is for avoiding the loss of the carbon in those trees. Put another way, little of this is supporting actual efforts to slow industries that currently spew carbon, nor getting carbon out of the air.

Carbon offset markets can still be an important incentive and tool, but I think we can all agree that there need to be high standards that minimize greenwashing and bullshit credits. The lack of consensus and regulation on those standards certainly undermines confidence in this market. Even though many leading companies have made public commitments to reduce and offset their emissions, most of them do not want to take on much risk by spending real dollars that will be perceived as greenwashing. For the serious actors in the space, this problem has to be remedied if we are going to see any real growth in the voluntary offset market.

Another major concern is that by focusing too much on the offset market, we are going to allow big emitters off the hook, avoiding pressure to decarbonize their operations and supply chains. Just buy a few credits and, voilà, the people are happy. But in reality, any opening given to companies to get out of lowering their own emissions will put us further off course from meeting our reduction targets. I am empathetic to that concern, but we have been taking that position for decades, and generally the economy is not decarbonizing. We need to allow the market to find the lowest-cost ways to take real carbon out of the atmosphere with high integrity. In some instances that will mean reducing solely from a company's supply chain; other times additional offset purchases will be necessary. The simple truth is that we will need to decarbonize with every resource we have.

Ultimately, just having the "trade" part of "cap and trade" isn't going to be enough. We need to have a cap-and-trade system, like what we had for $SO_2$. There need to be strong caps on what sectors are allowed to legally emit and a highly regulated offset market to go alongside those caps. That argument has been well articulated for decades now, but unfortunately given the dumpster fire that is the U.S. Congress, that kind of regulation is a fantasy for the foreseeable future.

Estimates range widely, but the cost of the transition to a regenerative system is in the ballpark of $350 to $450 billion each year for ten years. But let's look through the system to see who is going to pay for that.

Consumers are stretched. Particularly when coming out of a high inflationary period, they simply are not going to pay more for more sustainably sourced products.

Many will point to the food manufacturers and say they should foot the bill. While I agree with them, I don't think there is much reason to believe they will without significant consumer and regulatory pressure—neither of which currently exists. While these companies are quite profitable, they are generally high-volume, low-profit-margin businesses. They don't have endless amounts of excess cash to spend without hurting profitability and seeing their stock get pummeled on Wall Street—the quickest way to lose your job as the CEO of a publicly traded company. They simply are not going to spend the necessary cash without a justification of a strong return on their investment. (I am not defending the short-term-only, profit-driven motive; I am just trying to lay out the reality we face.)

The processors are in between the growers and the manufacturers, and they too are low-margin businesses that will only invest the capital needed to create the supply chains to delineate regeneratively produced ingredients from conventional products if there is a sufficient demand from their customers.

That leaves farmers. Over the years, farmers have lost a mind-boggling share of the money spent on food. According to the USDA, in the 1950s growers took home between 40 and 50 percent of the dollars spent on food in the United States. In 2022, that number was down to 14.5 percent. There simply is no room for growers to foot this bill en masse.

If cleaning up the air has to be paid for, I'd be happy to see the biggest *polluters* foot the bill, because they are the ones who can afford to pay the highest price and who have the most intense societal pressure because of their role in creating the problem. The higher the price paid per ton of carbon by the big emitters, the higher the financial upside for farmers. The bigger the financial upside for growers, the faster we will see them adopt regenerative practices. Indeed, the potential new revenue source has reordered how many of the big agriculture groups in Washington approach climate change and regenerative agriculture. Instead of the mere mention of climate change being perceived as a direct attack, as it was when we were in the White House, it is now an opportunity to improve farmer livelihoods. The issue of carbon and regenerative agriculture has the executives of highly conservative agriculture groups like the American Farm Bureau Federation working with the Environmental Defense Fund and the Nature Conservancy to find a way to speed the transition.

No doubt there are risks and downsides to emphasizing carbon sequestration when we might otherwise be remaking agriculture in a more comprehensively sustainable way. Indeed, there's a fair argument to be made that focusing too much on carbon sequestration at the expense of promoting full biodiversity will get us only so far in the long term. But if I learned anything in Washington, it's that trying to enact complex change all at once is a great way to stay stuck exactly where you are, making no progress at all. And carbon markets are a true and fundamental *revolution* in the modern capitalist system that dominates the globe.

Think of it this way: Even as I described the flaws in forestry-based carbon credits, they still represent the first time we are placing value on the forest we preserve, instead of the wood we extract. It's the first time we are valuing what we put into the soil instead of solely the crops we extract. It is a breakthrough that, if it became a standard part of the economic system, would change what we value and how we value it. It would change how the economy relates to nature. Right now, chopping down the forest to grow soybeans is a logical economic decision. We have to make it a logical economic decision to protect it instead. If we don't, we will continue to watch agriculture degrade, deplete, and outright destroy the natural ecosystems that sustain life on planet Earth.

I think of carbon sequestration as a gateway. We know how to do it. We are well on our way to figuring out how to quantify it. And if carbon is retained in the soil, a slew of benefits automatically follow: water retention, increased biodiversity, better soil health.

And technology like Loam's can unlock carbon sequestration at a level that can combine with carbon trading to make a game-changing difference.

Carbon markets still have kinks that need to be worked out (most glaringly, agreeing on exactly how to measure carbon retention), but they now function in the United States, China, Japan, Canada, and the European Union. By 2022 the total value of carbon-trading programs had reached nearly $800 billion globally for all credits and is predicted to continue to grow meaningfully. I can foresee a day when an Iowa farmer makes as much off carbon as they do off soybeans.

Inoculating seeds at scale with natural, carbon-fixing fungi presents just such an opportunity. If the practice spreads with anything remotely close to the rate of GMO-crop adaptation, reducing massive quantities of atmospheric greenhouse gas levels in the near term becomes an exciting possibility.

## SWITCHING GEARS

Agriculture's contribution to climate change and ecological deterioration would be so much lower if only cereal crops like corn and wheat, upon which humans depend for so many of our calories, behaved like soybeans, lentils, and other plants in the legume family. Legumes have the ability to partner with soil bacteria in a good deal for both parties: The bacteria are protected and fed by the plants, and in turn they help produce nitrogen, which the plants use for growth. Other crops don't enjoy this mutually beneficial relationship, which is why farmers apply fertilizers rich in usable nitrogen.

There is no shortage of nitrogen. The gas is all around us; it makes up something like 80 percent of the earth's atmosphere. But in its pure form, plants cannot use it. More than a century ago, two German scientists, Fritz Haber and Carl Bosch, solved that problem by exposing nitrogen and hydrogen to heat and high pressure to create ammonia, a compound that can be broken back down by plants into nitrogen (ammonia can also be used to produce high explosives, its original use).

Artificial fertilizer made by the Haber-Bosch method was a boon to farmers, dramatically increasing yields, and has become a cornerstone of global food production. It is also a disaster for the environment. Its production requires enormous amounts of natural gas, which contributes to climate change. And about half of the nitrogen fertilizer applied to fields gets washed away, polluting streams, rivers, and oceans, creating the dead zones you might have heard of. All in all, nitrogen fertilizer use produces 2 to 3 percent of all greenhouse gas emissions, totaling an estimated 2.2 billion metric tons a year.

For a hundred years, scientists have been on a futile search for fertilizer's holy grail, a method to get industrial quantities of nitrogen to crops that avoids the environmental and financial drawbacks of Haber-Bosch. Numerous companies in recent years have raised large amounts of money to attack this problem, but none have yet been able to produce enough nitrogen naturally to replace synthetic nitrogen in any meaningful way. Tim Schnabel, a Stanford bioengineer in his early thirties who heads a California company called Switch Bioworks, thinks the grail is within his reach. The company is developing systems that use naturally occurring soil microbes to create nitrogen fertilizer in large quantities that nonlegume crops can exploit.

Soil bacteria do provide some nitrogen, even to nonleguminous plants, just not nearly enough. To boost that process, scientists have engineered bacteria that supply more nitrogen to plants, but the microbes do so at their own expense and can't reproduce at an adequate rate to significantly offset the need for chemical fertilizers. The reason for this is that once the microbes release nitrogen, they create an environment where they can no longer reproduce and effectively kill themselves off. Schnabel and his team are engineering these microscopic powerhouses to reproduce at will for the first stage in their lives, without making nitrogen contributions to neighboring plants. Then, only after their population has been established, do they switch (hence the company name) and begin producing nitrogen—at seventy pounds per acre, compared with only about three pounds for "non-switched" bacteria.

By planting seeds inoculated with the Switch microbes (or applying them directly to the soil at time of seeding), farmers can get the same yields for two to three times less money than they would pay for traditional Haber-Bosch fertilizers. Because the switched microbes make nitrogen only at the rate plants absorb it, there is no pollution from runoff. And the process uses only one-tenth of the energy of traditional fertilizer manufacture. It is still early days, but the company has successfully demonstrated its system in the laboratory and has started trials in fields to see how the microbes perform in various regions and soils. Accomplishing breakthroughs in a lab is one thing; getting biological solutions to work in soils that are their own complex universe is another. But the game-changing potential is real, and Schnabel hopes to see a full rollout to farmers by 2028.

## DEEP THINKING

Perhaps it's because I'm a landlubber from the heartland, but I, like many, tend to overlook the huge potential of the ocean in combating climate change. Most of us have an inherent bias toward a land-based perspective on both the problems and the solutions. But we'd be missing maybe the biggest chance we have to change our trajectory. Seventy-one percent of the earth is covered by seawater. The oceans are currently absorbing 25 percent of $CO_2$ emissions globally—about the same as the totality of green plants on land. And it's possible we can help them do more.

Bren Smith is anything but a landlubber. And he has developed a system that could allow aquaculturists to profitably farm the oceans and sequester tons of carbon at the same time. Raised in a small coastal village in Newfoundland, he quit school at age fourteen and became a commercial fisherman, eventually plying his trade in waters around the world. Smith admits that like many who catch wild fish for a living, he worked on boats that destroyed entire ecosystems as they threw thousands of pounds of dead bycatch back in the ocean.

For the uninitiated, commercial fishing works, like all other professions, according to the market. And if the seafood-buying market wants to eat only a small handful of types of fish, only those fish are kept, stored, and processed. The rest of the fish in the net—often the majority of the catch—is tossed back overboard, usually dead. So, a huge net might turn up hundreds of valuable cod, say, but also be filled with a thousand pounds of other creatures. There's plenty to be said about the problems of our seafood economy, but even the most conscientious fishermen have a hard time changing the way they do things for the environment's sake.

Smith's career came to an abrupt end in 1992, not by his choice. After five hundred years or so of being a mainstay of the lifestyle and economy of what eventually became Newfoundland, cod populations crashed. Canada imposed a moratorium on all commercial cod fishing in an attempt to save the remaining fish, instantly putting thirty thousand fishers out of work, one of whom was Smith. He had learned the hard way that there would be no jobs and no food in a dead ocean—or planet.

Today, Smith is executive director and co-founder of GreenWave, a Connecticut-based organization that promotes what he calls "regenerative ocean farming." Founded in 2014, GreenWave advocates for a promising method of producing food and other agricultural products while—you guessed it—removing greenhouse gases from the atmosphere. Regenerative ocean farms raise mussels, oysters, and clams similarly to how they are raised in conventional aquaculture, the source of nearly 90 percent of the bivalves humans eat.

Like the products of conventional aquaculture, those reared in regenerative operations provide a source of income when they are sold as human food. But regenerative ocean farms differ from conventional ocean farms in two key ways. One is that each regenerative ocean farm raises a variety of animal species instead of concentrating on one, as is currently typical. Those species live at different levels of the water column—basically a way of saying that they live at different depths. Mussels, in sacks, dangle on ropes near the surface; oysters, in cages, suspend farther down; and clams, in cages, sit on the ocean floor, creating a vertical, three-dimensional farm that produces huge amounts of food in a small area. While I've focused much of this book on climate and carbon reduction, remember that the food crisis is also about making sure our planet can produce enough food for our massively growing global population. Harnessing the power of the ocean to farm more sustainable, delicious, quality protein in the form of shellfish is without a doubt one of the best solutions we have at our disposal.

One of the main benefits of shellfish production is that these tasty little creatures do not require feed. They are primarily filter feeders that feast on microscopic plankton, bacteria, and organic particles suspended in the water. This avoids one of the biggest pitfalls of fish farming, which is the need to catch massive amounts of fish to feed the farm-raised fish, decimating key food stocks for the ocean's food chain.

The other key difference is that in addition to bivalves, regenerative ocean farms raise a variety of species of seaweed (kelp in particular) on long ropes. A small portion of the kelp is eaten by humans (a too-small portion, its fans claim; just ask diners who pay a fortune at places like Noma, one of the very best restaurants in the world). But most of the kelp gets processed into fertilizer, biofuel, and cosmetics. Such "farms" require no fresh water, no fertilizers, and, importantly, no added feed—zero inputs—even as they sequester massive amounts of $CO_2$ from the atmosphere. Smith says that anyone who has a boat and $30,000 to invest can create a profitable regenerative ocean farm.

"Regenerative ocean farming really brings together everything in one simple system, growing a mix of different species," Smith explains. "Think of it as an underwater garden."

Seaweeds such as kelp alone capture about 200 million metric tons of carbon a year. The World Bank calculates that ocean farms covering only 5 percent of U.S. water would offset the emissions of twenty million automobiles. Although no one is predicting that they will occupy that much at any time in the foreseeable future, there is no doubt that kelp's carbon-capturing potential is huge.

Kelp and other so-called macroalgae have prodigious growth rates that make terrestrial plants look like pikers when it comes to carbon sequestration. Some aquatic seaweed species grow up to two feet *a day*—attaining lengths of nearly ninety feet at maturity. Coastal waters can sequester more than twenty times as much $CO_2$ per day as land forests, lending credence to Smith's claim that aggregations of kelp are the "rainforests of the sea." Better yet, when kelp expires, much of it sinks to the depths, where it can remain for centuries or even millions of years.

Kelp could also provide sustainable food to humans. To that end, Smith and other kelp advocates have attempted to develop a menu of products such as pickles, purees, and seasonings. Kelp packs a powerful umami punch, sought after by many chefs. Nutritionally, it's a superfood with more iron than beef, more protein than soybeans, and more calcium than milk, along with impressive amounts of healthy omega-3 fatty acids. It is rich in fiber, magnesium, potassium, and iodine and contains vitamins A, B, C, and E.

But, at least among North Americans, eating kelp has yet to catch on. (Kelp—the next kale?) In our defense, I have tasted a lot of bad kelp products, but I am confident chefs and entrepreneurs will ultimately find ways to make kelp taste great; it's already enjoyed in many traditional Asian cuisines. Smith thinks its best chance for broad acceptance among Western audiences is by playing a supporting role in cuisine, not occupying the center of the plate. He compares kelp's possible uses as an ingredient to those of soybeans. Of course there's tofu, which we all know about, but products made from soy find their way into breakfast cereals, crackers, prepared breaded foods, hot dogs, cold cuts, baby formula, plant-based meat replacements, to mention just a few hidden sources. It may come as a surprise that Americans, many of whom have never tasted tofu, consume an average of nearly *135 pounds of soy* each year. Smith sees no reason to prevent kelp from becoming the new soy, tucked away in numerous foods, without causing the environmental damage of vast monocropped, GMO soybean fields. "Soy, but less evil," he says.

Smith started with a single small farm near the Thimble Islands in Long Island Sound. Today, GreenWave has morphed into a nonprofit that supports a global network of nearly four thousand farms. The original Thimble Island Ocean Farm has become a floating classroom for novices. It provides training, marketing assistance, and financial aid. Smith's goal is to see the GreenWave family expand by 2030 to ten thousand farms, each producing nutritious food and other products with zero inputs while removing tons of $CO_2$ from the atmosphere.

As powerful as this system is, there is a long road ahead before it becomes a meaningful part of the solution. A number of challenges remain that will have to be overcome. First and foremost, the market

for shellfish is relatively small, and for kelp, even smaller! There simply isn't enough market demand for a production system to grow
rapidly. This is especially true for kelp, which is not a food that is
widely consumed. For this approach to work, kelp will need to become
a normalized ingredient in our diets. The first step in that process is
making it taste delicious and integrating it into a wide variety of
dishes and products. If the market remains small, the positive impact
of this approach will remain small.

The challenge of small scale creates another big barrier: cost. Do I
foresee a future where we are all eating large portions of kelp on a
daily basis? It's unlikely; even in cultures where it's enjoyed, it's not
usually a main ingredient in a meal. The clearest way to scale this
system is to turn kelp into various ingredients that can be used in a
wide range of products. To do so, these ingredients will have to compete with the derivatives of crops like soy: Think protein, oils,
emulsifiers, and so on. Soy, as an example, has a tremendous advantage of scale, which results in a very low cost of these ingredients. We
have invested in all the systems—from genetics, to planting and harvesting and processing—to make very cheap the creation of these
ingredients. Right now, it would be roughly fifteen to twenty times
more expensive for a company to replace soy protein with kelp protein in its products. I don't know a company on Earth that would do
that. It's a bit of a chicken-or-egg problem: Will the demand increase
to cause people to invest, or can investment happen to make the
product more appealing and more economically viable? Improving
the production and processing systems, and getting the volume of
these products significantly higher, will be critical to bringing down
the cost and expanding the opportunities where these more sustainable substitutes will make economic sense. It is possible to accomplish,
but a lot of work remains.

But it will be worth it. It's not going too far to say that if we maximize the carbon-capturing potential of the wet 71 percent of the planet as well as that of solid ground, we can have the ability to farm our way out of some of the worst effects of climate change. But to do so, we will need far more attention, investment, and resources dedicated to unlocking the ocean as a way to combat climate change.

## REGENERATIVE CAPITALISM?

So many of the incredible conceptual and technological advances I've written about here show the power of human ingenuity to harness science and the awesome powers of the natural world. But you might have noticed a consistent refrain in all these stories: How do they get paid for? How do we incentivize actors across our food and agricultural system to make a different set of choices, to improve the well-being of humans and the planet? That's a matter not of science or nature but of society and economy. This is the part of the picture that is *completely* man-made, so you'd figure we ought to be able to achieve some innovation there, too, right? No new tool or set of tools alone will be able to overcome how we currently value—or don't value—our natural systems. If we don't begin to shift our economy and society's relationship to the natural world, these new technologies will fall short of solving the challenges ahead. But here, too, there is hope.

I might not have bothered to take the trip to Douglas Eger's farm near Charlottesville, Virginia, had one of his staffers not matter-of-factly assured me over the phone, "He will blow your mind."

And he did.

When I arrived, Eger invited me to settle into a chair beside him on the expansive porch of his home, overlooking the rolling patchwork of central Virginia. Bordered by well-maintained wooden fences, the setting included serene herds of browsing cows and horses.

The morning was so pleasant that I had to remind myself I had come to see Eger to discuss climate change, the force that, if not reversed, could devastate this bucolic landscape and others like it everywhere.

Eger believes he has found one of the solutions to global warming: Harness the power of the marketplace, the thing that, despite our best efforts to control it, has gotten us into the unholy mess we face today. He claims that his new company, which has the enigmatic name Intrinsic Exchange Group (IEG), has developed the means to "tame the destructive beast" of the market.

Several years ago, while living in California's Sierra Nevada, he realized that the existing efforts of conservation groups and government policy makers to stop the destruction of natural systems were grossly insufficient. "I got tired of losing," he said.

His idea of losing requires considerable perspective. He and his wife, the sculptor and filmmaker Cristina Khuly, have certainly done more than their share to conserve natural resources. In the early 2000s they purchased nearly seven thousand wild acres in New York state's Catskill Mountains and then partnered with the Open Space Institute and the Trust for Public Land to permanently preserve the tract. Still, he knew that even protecting that vast expanse, with its verdant mountains, popular hiking trails, and pristine trout streams, was not going to make a dent in a global environmental problem. "Creating that little ecological island was satisfying, but it was a drop in the bucket against the needs of the world," he said.

As a young man, Eger was told by his father that if he truly learned how the forces of finance and business operated, he could use them to drive change for the better. In the mid-'00s, during his Sierra Nevada hikes, he recognized that the current economic system has a fundamental flaw. The market—which like it or not is the ultimate boss in our capitalist system—values only extraction when it comes to natural resources. But there is so much more that has value in nature, and the global economy has not priced the value of those additional "services," or, more accurately, we have thus far put the value more or less at zero.

He considered the example of a lumber corporation that owns a forest. Today, those woodlands' financial value is based on timber. To make money, the company has to cut the trees down, thereby destroying a significant source of the land's value. The forest itself is not valued beyond the current and future amount of wood it holds. The same is true for farmland and pasture: Their value comes from the crops or meat that can be extracted, the vast majority of which is produced in ways that deplete the soil.

The market puts no value whatsoever—*zip*—on anything else those fields and forests provide. Loggers get paid pennies per board foot of wood, period. But if a tree was seen as a fancy invention that stores carbon and keeps it out of the atmosphere, cools the climate, conserves water, prevents floods, provides wild animal habitat, and, along with a few hundred thousand of its brethren, offers recreational opportunities, it would be worth a lot while living. So, Eger's thought was, What if there was a financial vehicle that would harness the true value of a tree or swath of grassland without destroying it?

Eger, who is in his mid-sixties, has spent his work life thinking big and then acting to make his visions come to life as successful businesses. He's a cross between a serial entrepreneur with a Midas touch and a modern Renaissance man. He chaired the internet company Pet360 (which included PetMD.com and PetFoodDirect.com) until it was sold to PetSmart in 2014 for $130 million. He became chairman and CEO of Sheffield Pharmaceuticals, a Connecticut-based maker of over-the-counter medicines sold in drugstores nationwide. He founded TechSource, a firm that helped colleges and universities move intellectual property rights, such as the patents they hold, from pure research into useful and profitable practice. He was co-founder of Kachina Capital, a firm that advised companies on strategic and financial issues.

Not enough for one man? Well, he also had a separate career as a successful filmmaker, producing documentaries that received Academy Award nominations and won an Emmy. Clearly, Eger has the sort of eclectic background and the unconventional problem-solving approach needed for someone who intends to turn the economic system upside down and help protect and restore the environment in the process.

Eger suggested we take a walk across his farm. Their two beautiful dogs came along, leash-less but never straying far from Eger's heels and immediately obeying his every command. Periodically, he threw a ball, which they dutifully scrambled to fetch. I'd never encountered better-trained dogs. The purported reason for our walk was so that Eger could show me concrete examples of nature performing valu-able services. When he suggested that we wade across a thigh-deep, fast-flowing stream despite my wearing jeans and leather hiking shoes, I realized that he was testing my mettle. Was I seriously interested in his mission, or just another venture capital jerk taking up his time? So, into the cold stream we went.

As we hiked up through a forest, he explained that his solution to the problem of the current economic system was to create a completely new financial asset class called natural assets as a way to generate capital from nature. "Let's change the incentives," he said. "Now the logical way to get rich is through extraction. Why not assign value to carbon retention, flood prevention, and the ecosystem's other services?"

After all, they do have intrinsic value. In New Orleans, he pointed out, engineers have spent something like $14 billion to build levees to hold back floodwaters that natural mangrove forests—now lost to development because the market did not place a value on them—

once controlled at no cost whatsoever. The price of California water rights on the Nasdaq exchange soared to more than $1,400 per acre-foot in 2022, up 56 percent from the beginning of that year. Much of that water comes from snowpack in the mountains. Even bats, not often viewed as a valuable resource, thrive in undisturbed forests and gobble enough destructive insects to prevent more than $1 billion a year in damage to trees and crops.

If you take the value of all the services well-managed farmland and woodlots provide, put a price on them, and create a financial instrument that allows investors to make money from changes in that price, suddenly conserving an ecosystem becomes more profitable than ruining it. And the better job you do in growing those assets (say, managing a conventional farm with regenerative techniques to improve carbon and water retention), the more profit you stand to make.

We have *never* assigned economic value to these services, and that has had dire consequences. Take mangroves again, as an example. These are some of the most crucial ecosystems if we want vibrant, healthy oceans. Around the globe, mangroves are the hatcheries of countless species of fish that underpin the complex food web that feeds well over a billion people. Fewer mangroves, fewer fish. And while we have a price for the fish we ultimately catch, we are not valuing these natural hatcheries.

Say there is a developer looking to build an oceanfront hotel on land that has mangroves on it. In our current system, they would run their numbers and assess the value of what that land could generate, versus the price of the land and the associated costs of clearing the mangroves. There is no consideration of the value that mangrove has in terms of its production as a hatchery in the food economy, or as a buffer against flooding inland, nor other ecological services it provides. I would guess more often than not, that hatchery is more *economically valuable* than another hotel. But because we don't price that, we are unable to make that assessment, and the logical answer economically in this case is to cut the mangroves down. This plays out across the globe where economic development meets natural ecosystems.

To that end, Eger invented a new corporate entity he calls a natural asset company (NAC). Eger is working to make NACs publicly traded entities, but they can also be constructed privately. The idea is to take an ecosystem, structure it like a company, and assign value to the goods and services it is providing.

In a public market setting, investors would be able to trade a NAC's shares on exchanges, just as they buy and sell shares of any other corporation or commodity. Traders stand to reap capital gains by selling those shares for a higher price if the value of the NAC's preserved natural assets rises in the same way owners of stock profit when the value of a company's shares rises, or owners of commodity futures profit if the value of commodities rises. Owners of the NACs invest proceeds from stock listings in ways that increase the value of the natural assets in the land they control, creating even more intrinsic value.

As with other companies on the market, the value of a NAC would be based not only on what it is producing now but on its potential for future growth. A NAC can and often will include aspects that generate direct revenue while also accounting for and assigning value to other outcomes of that ecosystem. To give an example: A regional dairy co-op could create a NAC. The balance sheet for the value of the entity would account for how much revenue was generated from milk and cheese sales, like a normal business. But it would also assess the health of the soil, the amount of carbon being stored, improved water management, or increased biodiversity, for example. By establishing a NAC, the quality and vitality of that ecosystem can now have value, as opposed to just the product—the milk. The better maintained or more vibrant the ecosystem, the better and the more plentiful the milk could be as just one positive, profitable outcome, and the NAC is able to deliver on profits for shareholders.

The potential is huge. According to a 2014 study by the economist Robert Costanza, the world's natural assets provide services worth $125 trillion per year. That figure is far larger than the GDP (the total value of all goods and services produced) of all the countries in the world, which the World Bank estimates to be $90 trillion per year. Today's economic measures completely ignore the contributions of natural assets. As a result, the decisions we are making on a global basis are deeply distorted and inherently irrational. We are destroying things of greater value to society and replacing them with things that hold inherently less value.

At first, I found it difficult to get my head around Eger's ideas. They were utterly foreign to my understanding of how financial markets work. But as I listened to him, it became clear that NACs were far more than just Eger's pie-in-the-sky dreams. This amounted to a fundamentally more accurate way to assign economic value to the world around us. It would take the cornerstone of a carbon credit—one of the first transactions valuing an ecosystem service—and expand that dramatically to comprehensively and accurately value the natural world around us. This would transform how we made decisions as a species. And if NACs end up being listed at public companies, they would allow anyone with some savings to invest in a way that financially supports organizations that improve the environment, not just to enjoy feel-good emotions from doing the right thing, but to make real money.

He explained that to become listed on a stock exchange, a government, business, farmer, or tribe—any entity that controlled the right to exploit the natural assets of a defined area, be it forest, field, rangeland, or even the seabed—would first have to calculate the value of the assets it held. Eger has set the minimum value to qualify as a NAC at $200 million.

That sounded complicated; so many factors would have to be taken into consideration to actively, accurately assess the value of a NAC. But Eger pointed out that for more than two decades economists and public officials have developed sound methods of determining the value of various natural assets. IEG uses widely accepted criteria from such institutions as the UN's System of Environmental Economic Accounting, which use these methods to determine the amount of damage a hurricane has caused, for example, to arrive at its dollar valuations for NACs.

Eger also got advice from Robert Herz, a past chairman of the Financial Accounting Standards Board, the organization that establishes accounting and reporting standards for private companies and nonprofit organizations that follow "generally accepted accounting principles," a phrase found on most stock prospectuses and financial statements. With traditional publicly traded firms, quarterly and annual reports that adhere to these standards provide investors with a transparent and uniform way to determine the financial health of a business—its income, profit or loss, cash flow, debt, and the like. NACs will issue similar investor reports. But instead of traditional measurements, their reports will provide "relevant, reliable, and understandable information on the flows of the ecosystem services and their stocks of natural assets," according to Herz.

As a further testament to the soundness of Eger's NAC concept, the New York Stock Exchange has not only expressed an interest in listing shares of NACs but invested in IEG, as have the Inter-American Development Bank and the Rockefeller Foundation, a $6 billion philanthropic organization with a mission "to promote the well-being of humanity throughout the world."

IEG will make its money by helping interested groups to create a NAC by using accepted methods to determine the value of their assets and dealing with regulatory bodies in much the same way that investment banks help corporations looking to list shares on stock exchanges today. Eger has already been contacted by a flood of organizations, including corporations, indigenous groups, and national governments, interested in forming NACs. One major administrative hurdle facing Eger is gaining approval from the Securities and Exchange Commission (SEC) to sell NAC shares to the public. The SEC considered Eger's application, but the regulatory body is notoriously secretive about its deliberations, so no one can say when, or if, it will give its blessing. Eger remains hopeful. "Natural assets have the most fundamental of values," he says. "If we don't have a functioning ecosystem, we don't have a functioning economy. We may not even survive on the planet."

It's one thing to list a company's shares but entirely another to persuade the public to invest in those shares. "Is everyone going to jump into NACs? Probably not," Eger said. But he adds that investors have put their money into much less tangible assets than the trees, soil, and water held by NACs. Bitcoin and other cryptocurrencies are worth billions of dollars, and their valuations are based on nothing more than computer algorithms. Collectors paid real money for non-fungible tokens representing digital artworks, songs, and videos. One website, CryptoKitties, allowed people to use cryptocurrency to "buy, feed, and breed" limited-edition digital cats. On another, investors could buy and sell "real estate" located only in the realm of cyberspace.

A more accurate comparison with NACs, according to Eger, might be shares of a company like Warren Buffett's Berkshire Hathaway, which pays no dividends, even after a history of steady profit and growth. The only way investors can make money from shares in the hugely successful firm is to buy stock and hold it in anticipation of the company growing and generating higher profits, which will push the value of its shares upward. Investors do not make a cent until they sell the shares. Buyers of shares of NACs stand to profit similarly.

Eger points to the eager consumer and institutional interest in making so-called green investments in companies that support environmentally friendly products and services. In 2021, investors had put more than $500 billion into so-called ESG (environmental, social, and governance) funds. But the problem there is that there is no way for investors to ascertain exactly what they are buying into and to determine the accuracy of organizations' green claims. "Consumers want to do the right thing, but they want to do so in ways that are simple," said Eger. "Today green investors have to worry about everything. Is this or that claim real? It becomes overwhelming. You almost need a PhD to figure it out. NACs would make green investors' lives simpler by providing regular reports with hard data showing how a company benefits the environment."

Back on the porch, Eger and Cristina served me a delicious lunch of salmon, salad, and fresh green beans. We said our goodbyes, and I walked to my car, covered in mud and still wearing damp pants and muddy shoes, thankful for the clean clothes in my suitcase. My mind churned. Would Eger succeed? I hoped so but couldn't be sure. I could envision a world with hundreds of publicly traded NACs, protecting millions of acres and being worth billions of dollars. At the same time, NACs would also provide anybody with even modest savings a vehicle to make investments that would allow them to directly support businesses that contribute to the lowering of greenhouse gas emissions. A huge upside.

Eger's an environmentalist, no doubt about that, but he also sees a possible way to make tons of money—and more important for the rest of us, giving the market a profit motive to save vital lands and ecosystems from shortsighted development.

Today, there are many different groups working on valuing nature and its ecosystems services differently. We have known about the problems of environmental degradation for decades, and despite some successful efforts to conserve and protect, the power of the global economy is just too strong. If we don't find ways to balance the relationship of capitalism and the environment, economic pressures driven by our current perverse incentives will continue to overwhelm whatever protections are put in place. But if people saw more value in protecting and restoring than in extracting and destroying, we could see the power of the market unleashed to help solve the most existential threat humanity has ever faced. Capitalism has worked against our food system, our health, and our climate for so long. It's time to make it work *for* those things.

# Conclusion:
# A Fork in the Road

## FARMER BROWN

Can we adapt the food system in time to avoid catastrophe? I have to be honest. With the efforts of companies like Inari, Loam, GreenWave, and IEG, I'm truly hopeful, but I can't say that I'm *optimistic*.

Frankly, it's impossible to be optimistic that we'll avoid *any* kind of disaster when I take a sober look at what the science tells us is happening to our climate, and what anyone with eyes and a functioning brain *knows* is happening. Changes that were once theoretical or occurring out of sight are now painfully evident. Everywhere, farmers and ranchers are struggling to produce food under increasingly apocalyptic conditions: hurricanes, tornadoes, floods, droughts, wildfires.

For decades, climate models predicted these events. It turns out they were dead wrong. They *underestimated* the pace at which the world would be impacted by the changing climate. So, yeah, some days I'm pessimistic that we'll get out of having to deal with some hardcore disasters. But how many, what kind, and how often these disasters happen can make all the life-and-death difference in the world, which is why an all-hands-on-deck approach matters. At this point we can't fully stop climate change, but we need to do everything we can to mitigate it and slow it down and prepare ourselves to live and thrive under very different conditions.

So, what leaves me with room for hope? Well, let me tell you a story about Gabe Brown. He farms five thousand acres just east of Bismarck, North Dakota, purchased from his wife's parents in the early 1990s. For a few years, the couple continued to manage the massive ranch according to the conventional methods that Brown, like most other young midwestern farmers, had learned in college: abundant applications of chemical fertilizers and pesticides, reliance on a few cash crops, and raising big, fast-growing cattle that required a corn-heavy diet.

He nearly lost the farm.

Brown Ranch suffered a series of unforeseen disasters of near-biblical proportions. In 1995 and 1996, powerful hailstorms flattened the Browns' entire corn crop. His expectation that sales from their cattle herd would offset those losses were dashed a year later when a blizzard killed many of the animals. A summer of record drought followed yet another devastating hailstorm. Crops withered in the fields. Most nearby farms were hit by one or maybe two of these catastrophes. The Browns' farm was clobbered by every single one. To survive, Brown had to make the tough financial decisions typical of any business facing bankruptcy, and soon entered a downward vortex toward financial ruin. He skipped paying the principal on his loans, staving off foreclosure by making the minimum amount of interest payments. He sold cows and calves that in less desperate times he would have kept to build and replenish his herd—depleting the very resource that was keeping Brown Ranch hanging on. He skipped buying crop insurance that might have protected him from some of his losses. Both he and his wife had to take off-farm jobs. Those and ranch duties meant that they worked every waking hour, and many that they should have spent sleeping. It got so bad that he couldn't afford baling twine to bring in the needed hay for his animals and had to let the crop rot in the fields. More fortunate neighbors rubbed their hands in anticipation of picking up his acreage at fire-sale prices.

Desperate, Brown approached his banker, cap in hand, all but begging for additional credit that would enable him to buy the seeds and chemical inputs that he thought were his only chance of pulling through one more season. The banker turned him down flat.

Brown now says that the rejection was the best thing that ever happened to his ranch. Too broke to buy synthetic fertilizers and pesticides, he began researching the history of farming to find out how agrarians raised food before the advent of modern agriculture and its expensive inputs. The answer was regenerative farming.

Brown stopped tilling his land and gave up applying synthetic chemicals, both of which kill soil microbes and wreck its structure—and both of which cost money. He sold off the implements he used to plow and spread those chemicals, and used the proceeds to buy a seeding drill that let him plant crops on untilled land. Instead of depending on just a few crops, he sowed wheat, triticale (a wheat/rye hybrid), oats, corn, sunflowers, peas, hairy vetch, and alfalfa. The result: He no longer cared whether his corn failed or corn prices became dismally low because he could rely on income from plenty of other crops. The resulting biodiversity attracted pollinating insects and beneficial soil organisms that thrived once they were no longer bathed in chemicals. Worms! He saw worms on his land, and mused that before he couldn't have dug enough on the entire ranch to go fishing. These changes began to happen in a single season.

While formerly he left his ground bare during the winter, exposing it to erosion, Brown started planting cover crops that included warm- and cool-season plants, broad-leaved plants and grasses, plants with fibrous roots and those with thick taproots: pearl millet, sorghum, buckwheat, hemp, radishes, ryegrass, turnips, soybeans, sugar beets, clover, rapeseed, lentils, and mung beans. When left to decay, they provided the land with a balanced "diet" of nutrients.

His rangeland eventually supported a bovine banquet of more than a hundred different species of grasses, broad-leaved plants, and shrubs. Brown switched over to a smaller cattle variety bred to thrive on forage (as opposed to breeds meant to be fattened on corn in feedlots). To imitate the habits of wild ruminants, he kept his herds tightly bunched on as little as an acre or two at a time, even though they could have roamed over thousands. He moved them frequently, sometimes once each day, only returning them to a paddock after it had had several months to regrow. The animals trampled weeds, and their excrement fertilized the soil. The land became more fecund with each cycle, which allowed it to support more animals, which in turn made it even more fecund. The vortex of doom became an upward spiral.

He added laying hens to the ranch. They came onto the rangeland after the cows had moved on, just as small birds follow herds of wild ungulates in nature. Brown augmented the food the chickens scratched from the pasture with unsalable grain he'd grown that had been damaged in processing and would otherwise have been thrown out. The flock not only provided Brown with another revenue stream in the form of eggs and chickens but consumed insect pests. Sheep and hogs rounded out the livestock population and became two more sources of income. Brown wagered that diversity would give his ranch the resiliency to survive whatever nature threw his way.

And it worked. His cropland now yields 20 percent more than average for the area. It sequesters ten times as much carbon. And at a time when the same neighbors who once salivated over the prospect of buying his bankrupt farm for a pittance are delighted to get a few dollars of profit per acre, he's making $100 per acre, allowing him to purchase and rent more land. A bigger, more productive tract enabled his son Paul to join as a partner at Brown Ranch after graduating from North Dakota State University. He now manages the day-to-day operations of the ranch, assuring that a third generation will continue the family tradition on the same property.

There are a few key points to keep in mind about Brown Ranch, and they are what make me hopeful. One is that it is Big Ag in every sense of the word. It's not a sweet little boutique farm that sells solely at a farmers market in a well-to-do neighborhood. It is situated in the heartland of industrial farming. Fully one-third of U.S. farmland is in the Great Plains and Midwest. And prior to Brown's changes, it was operated under the exact same principles that make large-scale conventional agriculture so damaging to the environment—use chemicals and concentrate on a few crops. Brown completely turned around his operation. With the right incentives and supports to lower the risk and increase farmer upside as we previously discussed, virtually every one of the more than 700,000 farmers in the central states could replicate his results. Now imagine that thousands of those growers adopt regenerative practices, some with financial support from the likes of PepsiCo, Unilever, and other big corporate buyers who see such partnerships as a way of guaranteeing the security of their supply chains in the era of climate uncertainty. Many might be planting seeds from companies such as Inari and Loam, thereby using less water and fertilizer and pumping vast quantities of carbon out of the atmosphere and into the soil. They will be profiting nicely by selling that carbon on well-regulated exchanges. That would be the very definition of an upward spiral. And this stuff already exists and is being used on real farms.

In making the changes he did, Brown not only saved his own ass but did the rest of us a favor. The diversity of crops he planted on the formerly mono-cropped land and his abandonment of tilling sequestered tons of carbon in his soil, preventing it from escaping into the atmosphere as carbon dioxide. More to the point, he performed this vital service for no compensation from us whatsoever.

Conventional farmers are understandably conservative, in the small-c, literal sense. They pay dearly for inputs such as chemicals, fuel, and seeds and carry crushing debt loads on the machinery needed to tend hundreds or thousands of acres. Even in good years, they have no room for financial mistakes. In bad years, they have to do everything in their power to minimize their losses. And, given the ever-decreasing number of farmers in our country, oftentimes their best efforts are not enough for them to remain financially afloat.

A tragedy, to be sure, but even that could have a silver lining. Ray Archuleta, a Missouri-based soil scientist who has taught operators from across North America how to farm regeneratively, says that the majority of producers who adopt these environmentally friendly methods find themselves in positions similar to Brown's. Currently the farmers who are using regenerative practices generally either are visionary producers or were forced to because the alternative is insolvency.

Regenerative farming is practiced on only 1.5 percent of arable U.S. land, but at the current growth rate that will quadruple over the next decade—still a small portion, but given the size of American agriculture, enough to have a significant impact. Archuleta offers one explanation for this expansion. He has found that once farmers move to regenerative practices, they almost never go back. Gabe Brown's experience explains why. With the right cultural, economic, and policy shift, we could see this number increase far more rapidly than the trajectory we are currently on.

Given the volatility that lies ahead, the lack of diversity is a major weak spot in our food system. Today's industrial system, with its endless mono-cropped fields and vast livestock confinements, represents the logical culmination of a series of decisions made during one of the most benign eras of climate stability in recorded history. This stability allowed us the luxury of taking the risks inherent in putting all of our eggs into far too few metaphorical baskets. We are eating only a few plants and animals and, of those, a historically small number of varieties and breeds. We have lost an astounding amount of genetic diversity in our food. But the era of a stable climate with abundant resources has ended. We don't know enough about the challenges of a hotter, more unpredictable time, other than that they are coming and they are going to be big.

Diversity is our best defense: a much greater variety of crops; farms of all sizes; local foodscapes as well as national and international distribution systems; greater variation in the gender, age, and ethnicity of farmers and our political leaders; and the varieties and genetics of the plants and animals we raise. These different plants, animals, and people will help us gain resilience and give us ways to navigate the complexity of the future. The different approaches lower our risk, with some plants and animals doing better in certain conditions while others don't. Here's where buying from small and local farms can play a role; smaller operations are core to preserving and increasing the diversity in our production system.

In the end, there are no shortcuts to real and lasting change. We have to keep pushing the cultural transformation that will underpin the political and business changes we need. This is where all of us come in. Asking questions, raising concerns everywhere we live, work, eat, and sleep, is the task for us all. Have conversations at your jobs, places of worship, PTAs, neighborhood councils, state and local governments, in your home, on your text chats. Raise the questions about what we are eating, how it's produced, and the impact on people and the planet. It may seem out of touch when people are already challenged by the cost of food; the rejoinder is that the cost and the availability of food are only going to go in much worse directions if we don't take steps to fix the systems now.

As the culture continues to shift, we will have the opportunity to translate that shift in values into political and market-based gains. We will find moments where political stars align and we have the opportunity to get some major changes through. There will be major opportunities for us to clean up what companies are putting in our food and create the conditions that incentivize growers to farm in a far more regenerative way. This is critical. If we can harness the power of consumers in a way we simply have not to date, we *will* be able to build the companies of tomorrow, reward the companies of today that meaningfully change, and punish companies that continue to produce food in ways that are destructive.

Without these shifts, new technologies and tools alone will not be enough to solve our challenges. The companies described in this book, like Loam, Nbryo, and Intrinsic Exchange, are examples of the kinds of new approaches we need. Some of them may change our lives; some of them will certainly not make it. We will need our best minds to continue to develop new ways of food production to be able to nourish a growing population in a way that does not exceed our planetary boundaries.

All of this is possible, but none of it will happen overnight. We have made real progress over the last couple decades in some respects, but we have to find ways to make far faster progress in the years ahead, or the change will be forced upon us and not on our terms. At stake is our ability to hand down to future generations the quality of life we have been so lucky to have. The only path forward is with conviction and resolve, working every day to make better choices in our own lives and focusing to bring the core values of better, healthier, more regenerative food to the fore. Our way of life, our food, and our planet depend on it; they depend on us.

## THE NEXT SUPPER

Shortly after I prepared my Last Supper in Davos, I got invited back to cook another, a significantly different meal from the one I prepared with the help of the rotund, Swiss-German chef I'd so embarrassingly underestimated. This time, the World Food Programme sponsored the event, and the 110 attendees came from a lofty demographic made up of corporate giants, political leaders, and even royalty (I had no idea there were still so many queens and kings around). My second Davos dinner, which I called A Fork in the Road, was meant to demonstrate steps that all of us can, and must, take immediately in the face of climate change.

I prepared two meals that the staff served simultaneously, arbitrarily giving half the gathering one of the meals, half the other. The first group got a marbled, twelve-ounce, corn-fed rib eye, chargrilled and served with roasted potatoes and spinach sautéed in butter. For the second group, I prepared a fragrant stew of plump white beans that I had simmered in chicken stock along with carrots, onions, tarragon, and bay leaves. Once they had cooked, I removed half of the beans and pureed them, adding the creamy results back into the pot. Using the bean mixture as a foundation on the plates, I added ultra-fresh spinach that was slightly warm and crowned that with a few slices of meat from young, free-range chickens. The plate was finished with a salsa verde made from parsley, cilantro, red wine, and olive oil. As a final step, I sprinkled pistachios on top. Immediately, I noticed guests murmuring and bargaining with each other to swap plates.

At the end of the meal, I stood and said, "Okay, raise your hand if you had the plant-forward meal—the beans with a little chicken."

As expected, half of the guests' hands shot into the air. I went through some back-of-the-envelope math on the greenhouse gases and water used to serve the fifty-five people who received this plate.

"Now, raise your hand if you had the steak," I said.

Not a single hand went up.

"Come on," I said. "Exactly half the plates that left the kitchen had steaks on them. I got served the steak myself!" The other half of the diners tentatively raised their hands.

To that group, I said, "According to the Food and Agriculture Organization, the meat-forward meal I just fed you contributed nearly four times—*four times*—as much greenhouse gas to the atmosphere as the plant-forward dinner your companions ate." The amount of water used was equally alarming.

As the chef, of course I had tasted both versions of the entrées served at the Fork in the Road banquet. And here's the truth. No one loves a great steak more than I do, and my Davos steak was everything a scrumptious steak should be: caramelized brown on the outside with a deep pink interior, and each bite imbued with tangy richness. But a meal where animal protein played only a bit part—really not much more than a seasoning—allowing beans and herbs to shine, was far more satisfying than the steak in every way: filling, totally craveable, extravagant, even fun. The well-fed dignitaries agreed. The future can be, and has to be, delicious.

"We are at a crossroads," I said. "We have a real choice to make. What we put on our plates is having a huge impact on climate change. We have to change, and that change must start not just in banquet halls like this or in restaurants but at every kitchen table. And we don't have to give up pleasure."

Then I posed a simple question: "So, which plate is it going to be?"

# Acknowledgments

First and foremost, my deepest thanks go to **Barry Estabrook,** whose collaboration and dedication helped bring this book to life. Barry is a brilliant writer, and this book simply wouldn't exist without him.

To **Suzanne Gluck,** thank you for your steadfast representation and for helping make this book possible from the very beginning.

I'm grateful to **Francis Lam,** not only for his sharp editorial guidance but for his friendship throughout this journey. I'm also indebted to the incredible team at **Crown** for their unwavering support and belief in this project.

To everyone mentioned—and many others—thank you for your patience, encouragement, and grace during the many more years than it should have taken to bring this book into the world.

To my **Acre family,** and to the many advocates, farmers, entrepreneurs, policy makers, and eaters whose voices and actions helped shape this book—your contributions are at the heart of this work.

And finally, to **Cy and Rafa Kass,** who inspire me every day to keep fighting for a delicious, just, and bountiful future.

# Index

# ABOUT THE AUTHOR

Sam Kass was senior policy adviser for nutrition policy in the Obama Administration and is currently an investor in several food technology start-ups. One of Michelle Obama's longest-serving advisers, Kass was the executive director of her Let's Move! initiative and helped create the first major vegetable garden at the White House since Eleanor Roosevelt's Victory Garden. He is a graduate of the University of Chicago and was trained by one of Austria's greatest chefs, Christian Domschitz.